SpringerBriefs in Physics

More information about this series at http://www.springer.com/series/8902

Suzairi Daud · Jalil Ali

Fibre Bragg Grating and No-Core Fibre Sensors

 Springer

Suzairi Daud
Laser Center, Ibnu Sina Institute for
    Scientific & Industrial Research
Universiti Teknologi Malaysia
Johor Bahru, Johor
Malaysia

Jalil Ali
Department of Physics, Faculty of Science
Universiti Teknologi Malaysia
Johor Bahru, Johor
Malaysia

ISSN 2191-5423   ISSN 2191-5431   (electronic)
SpringerBriefs in Physics
ISBN 978-3-319-90462-7   ISBN 978-3-319-90463-4   (eBook)
https://doi.org/10.1007/978-3-319-90463-4

Library of Congress Control Number: 2018939013

Printed on acid-free paper

This Springer imprint is published by the registered company Springer International Publishing AG part of Springer Nature
The registered company address is: Gewerbestrasse 11, 6330 Cham, Switzerland

Preface

Alhamdulillah… Praised be to Allah s.w.t… Peace and Blessing be to Prophet Muhammad s.a.w…

While preparing this book, we were in contact with numerous researchers, academicians, and professionals. They have contributed toward the understanding and thoughts of physics and fibre sensors. In particular, we wish to express our sincere appreciation and gratitude to Prof. Dr. Jalil bin Ali from Faculty of Science UTM, Assoc. Prof. Dr. Hazri bin Bakhtiar from Laser Centre UTM, Nurul Amirah bt Yusof from Ministry of Health Malaysia, Muhammad Hakeem Anaqi, and Muhammad Hazeeq Arfan for their motivation, support, and patience. Last but not least, we like to thank our family members and friends for their endless support. Without their continued support and interest, the completion of this book would definitely impossible and keep in imaginary nature.

This book serves to design, construct, and analyze the optical FBG and NCF sensing system in indoor and outdoor environment, within different refractive indexes, under different solutions, and different temperatures. The FBG and NCF sensing system has been analyzed, the performance has been investigated, and the characteristics of the system have been reported. The performance of the system has been investigated in terms of Bragg wavelength shift, no-core wavelength, FWHM, and intensity. The fundamentals of FBG and NCF have been discussed in details. At the end, the model design of FBG and NCF sensing system has been developed successfully.

This book consists of six chapters, namely Introduction, Operational Principles of Fibre Bragg Grating and No-Core Fibre, Theory, Methodology, Research Finding, and Conclusion. The overview and motivations of this book are discussed in Chap. 1, which includes its objective. The principles of FBG and NCF are discussed in Chap. 2, which includes the writing techniques and preparation techniques of FBG and NCF. The historical perspective of FBG and NCF is also discussed in this chapter. In Chap. 3, the theoretical part of FBG and NCF is given. This includes the coupled mode theory analysis, operational principles of FBG and NCF, and characteristics of both FBG and NCF. Chapter 4 explains the experimental set-up and methodology used in the experiments, in order to prepare this book. Results, analysis, discussion, and evaluation of the performance of FBG and NCF sensing system are discussed in Chap. 5. Bragg wavelength shift and no-core wavelength are among the main criteria discussed in this book. At the end, the conclusions of this work are described in Chap. 6.

Hopefully, this book may lead to the fundamental study on FBG and NCF, especially in sensing field.

Johor Bahru, Malaysia   Suzairi Daud
                                                                              Jalil Ali

Acknowledgements

The author wishes to extend the highest appreciation to Laser Center and Department of Physics, Faculty of Science, Universiti Teknologi Malaysia, for providing the research facilities toward the success of this project. Sincere appreciation also extends to all beloved friends and families who have provided assistance in various occasions, motivation, and supports. Without their continued support and interest, the completion of book would have been definitely impossible. Thank you very much.

Contents

·

About the Authors

Dr. Suzairi Daud received his Ph.D., M.Sc., and B.Sc. degree in Physics from Universiti Teknologi Malaysia. He is currently a Senior Lecturer at the Laser Center, Ibnu Sina Institute for Scientific & Industrial Research and Department of Physics, Universiti Teknologi Malaysia. He has authored/co-authored numbers of technical papers published in international journals, a research book, and four-book chapter. His research interests include fibre Bragg grating, fibre sensor, fibre laser, optical solitons, nonlinear optical communication, and nano waveguides. He is a member of IEEE Asia Region, Malaysian Institute of Physics, and Malaysia Optics & Laser Technology Society.

Prof. Dr. Jalil Ali received his Ph.D. in Plasma Physics from Universiti Teknologi Malaysia in 1990. At present, he is affiliated with Laser Center, Ibnu Sina Institute for Scientific & Industrial Research and Physics Department, Universiti Teknologi Malaysia as Professor of Photonics. During 1987–2010, he has held numerous faculty and research positions including the Dean/Director, Bureau of Innovation and Consultancy. He has authored/co-authored more than 300 research papers for international journals, 11 books, and a number of book chapters. His areas of interests are photonics, optical solitons, fibre couplers, nano waveguides, fusion energy, and so on. He is UTM's delegate member for African-Asian Association for Plasma Training (AAAPT). He also holds memberships of OSA, SPIE, and the Malaysian Institute of Physics.

Abbreviations

CMT	Coupled mode theory
DNA	Deoksiribonukleid acid
EMI	Electromagnetic interference
FBG	Fibre Bragg grating
FBGs	Fibre Bragg gratings
FOS	Fibre optic sensor
FOSs	Fibre optic sensors
FWHM	Full width at half maximum
HiBi	High birefringent
NA	Numerical aperture
NCF	No-core fibre
OSA	Optical spectrum analyzer
TLS	Tunable laser source

Symbols

a	Core radius
b	Wavelength position at 0 °C
B	Magnetic field
d	Power dip
dB	Decibel
E	Electric field
k	Propagating constant
$\vec{k}$	Propagating constant vector
$\vec{k}_1$	Modal wavevector of the forward-propagating wave
$\vec{k}_2$	Modal wavevector of the backward-propagating wave
$\vec{k}$	Grating momentum vector
L	Grating length
MHz	Mega Hertz
M_p	Fraction of fibre mode power
N	Number of grating plane
nm	nanometer
n_{cl}	Cladding refractive index
n_{co}	Core average index
n_{eff}	Effective refractive index
n_o	Average refractive index
n_1	Refractive index of fibre core
n_2	Refractive index of fibre cladding
$pm/°C$	picometer per degree Celsius
R	Reflectivity
$R(L,\lambda)$	Reflectivity in the function of length and wavelength
s	Fringe visibility of the index change
T	Temperature
z	Distance along the fibre longitudinal axis
α	Thermo-expansion coefficient
ξ	Thermo-optic coefficient

ε	Permittivity
μ	Permeability
ε_z	Strain
Ω	Coupling coefficient
Λ	Grating period
Λ_g	Grating spacing
Λ_{pm}	Phase mask period
λ	Wavelength
λ_B	Bragg wavelength
λ_{in}	Incident light
$\lambda_{B,0}$	Nominal Bragg wavelength
λ_{NCF}	NCF wavelength
λ_o	Initial wavelength
δL	Change of length
δn_{eff}	Change of refractive index
Δk	Detuning wavevector
Δn	Dept of index modulation
ΔT	Temperature change
$\Delta \lambda_B$	Bragg wavelength shift
$\Delta \lambda_{NCF}$	NCF wavelength shift
°C	Degree Celsius
°F	Degree Fahrenheit
%	Percentage

Abstract

A prototype of fibre Bragg grating (FBG) and no-core fibre (NCF) sensing system has been designed, constructed, developed, and evaluated. Indoor and outdoor FBG temperature sensor system has been developed where the commercial FBG has been tested under controlled environment (lab based) and uncontrolled environment (outdoor based). The sensitivity of the system has been evaluated in different refractive indexes of solutions, different placements of sensor, under focused and unfocused elements and under different temperatures. The NCF has been incorporated into the FBG and the performance of the system has been evaluated. The combination of FBG and NCF can be used as a very useful tracking system, especially for the temperature and surrounding refractive index (SRI) sensing. For the purposes, TLS was used as the broadband light source and the output spectra been displayed through the OSA, where both transmission and reflection spectra of the system have been analyzed. The performance of FBG and NCF has been investigated in terms of Bragg wavelength shift, no-core wavelength, FWHM, and power dip. The sensitivity of FBG and NCF were calculated using the related formula, based on the data taken. Results show that the sensitivity of FBG and NCF system is directly proportional to the temperature change and SRI number, for both indoor and outdoor environment. Thus, a prototype FBG and NCF sensing system has been developed successfully.

Chapter 1
Introduction

1.1 Overview

This book will give an overview of fibre Bragg grating (FBG) and no-core fibre (NCF) sensors. The working principle of FBG and NCF will be discussed and explained in detail where the main motivation of writing this book is as a reference for researchers and academicians to gain knowledge on FBG and NCF-based fibre sensors. This book will focus on one of the basic SI (International System of Units) systems, which is temperature. Temperature is the most measured physical parameter for most processes in nature and temperature-dependent industries. Accuracy in temperature measurement or investigation is essential not only for the safety, but also for the efficacy of the operation and regulation of the industrial processes.

A sensor is a device that measures a physical quantity and converts it into a signal which can be read by an instrument. For example, mercury in glass thermometer converts the measured temperature into expansion and contraction of the liquid which can be read on a calibrated glass tube. A thermocouple converts the temperature measurement to an output voltage which can be read by a voltmeter. Temperature is an important and most commonly measured parameter in everyday applications. Traditionally, semiconductor sensors, platinum-resistance sensors, thermistors, and thermocouples are most commonly utilized for temperature measurements. For most applications, these conventional temperature sensors are adequate for their purposes. However, these all are not reliable due to the lack of intrinsic safety, their characteristics of being electrically active, and poor lifetime at excessive temperatures. These conventional temperature sensors are all point-based sensors. In other words, they are localized sensors, which can only provide a temperature reading over a small area rather than providing an overall temperature profile. These types of sensors are suitable for passive multiplexing, but difficult in practice, due to size limitations.

© The Author(s) 2018   1
S. Daud and J. Ali, *Fibre Bragg Grating and No-Core Fibre Sensors*,
SpringerBriefs in Physics, https://doi.org/10.1007/978-3-319-90463-4_1

Fibre optic sensor (FOS) has emerged as a modern device in sensing technology since the photonic probe made its tentative appearance in the mid of 1960s. FOS can be classified as fluorescent/spectrally based, intensity-based, or interferometric sensors. Fluorescent-decay, blackbody radiation, Fabry-Perot, interferometric, polarimetric, and dual mode temperature sensors are examples of sensors used in fibre optic temperature sensing. Due to its unique advantages, optical fibre sensor offers a number of distinguished and excellent advantages over conventional sensors.

The most important innovation of the FOS is the development of fibre Bragg grating (FBG) sensor. FBG temperature sensors are an ideal, intelligent distributed temperature sensor for real-time temperature monitoring. The principle of the FBG temperature sensor is based on the measurement of the reflected Bragg wavelength. FBGs are compact sensing elements, which are relatively inexpensive to produce, easy to multiplex, and applicable to a range of physical measurements. The gratings of FBG are uniformly spaced regions in the fibre where the refractive index has been raised from the rest of the core. These radiations scatter light by the Bragg effect.

The first fibre photosensitivity in germanium-doped silica fibre was discovered at Communication Research Center in Canada in an experiment conducted by Hill and his co-workers. Intense visible light from an argon ion laser was launched into the core of the fibre to study the nonlinear effects. The fibre attenuation was increased under prolonged exposure and the intensity of light back-reflected increased with time with all of the incident radiation back-reflecting out of the fibre. The increase in reflectivity resulted in permanent refractive index grating called Hill gratings. The experiment obtained a permanent index grating with 90% reflectivity for argon laser-writing wavelength. The modulation index change, Δn, was estimated to be approximately 10^{-5} to 10^{-6} within a very narrow bandwidth (20 MHz). However, the grating only functions at the writing wavelength and in the visible part of the spectrum. Their potential for applications in sensing and telecommunication will be recognized in view of the writing characteristics.

However, most of the recognized pioneering works about the FBG and its applications were only published a decade later after its discovery by a group of researchers at United Technology Research Centre. FBG plays a crucial role in sensing technology due to its unique smart structure. For this reason, it is a requisite for researchers to establish a better understanding of the FBG sensor in temperature sensing. Since the FBG's first discovery in 1978, numerous works have been conducted by researchers for the development of FBG sensor. FBG is one example of distributed Bragg reflector constructed in a short segment of optical fibre that reflects a particular wavelength of light and transmits all the others. This is achieved by appending a periodic variation to the refractive index of the fibre core, which generates a wavelength of a specific dielectric mirror. It can be used as an inline optical filter to block certain wavelengths or as a wavelength-specific reflector. Until today, FBG sensor is one of the most important, useful, and economical optical fibre sensors available.

Currently, FBG has been used as a very crucial sensor due to its excellent performance and sensitivity. As a result of its smart structure, excellent linear characteristics, immunity to electromagnetic interference (EMI), low fibre loss, and other outstanding advantages, FBG emerges as an important technology in fibre optic sensing. FBG is used in accurate measurement of temperature in a variety of environments, including harsh environment, underground, underwater, and in disaster areas. FBG sensors are dielectric and virtually immune to EMI. It can withstand hostile environments, including environments with high and excessive temperatures. It is capable of measuring temperature up to 1000 °C.

Multimode optical fibre is a type of optical fibre used mostly for communication over short distances, such as within a building or on a campus. Typical multimode links have data rates of 10 Mbit/s to 10 Gbit/s over link lengths of up to 600 m (2000 feet). For this reason, a little thinner no-core fibre (NCF) is used instead of traditional multimode fibre. The advantages of thin NCF is that it exhibits a more sensitive response to the changes in the environment. The jointing point with single-mode fibre (SMF) is a taper, which is easier for light to couple into NCF and will improve the sensing sensitivity. When the strain or curvature of the fibre changes, there will be a wavelength shift response of the sensor. Because there is no cladding, the NCF sensor is more sensitive than traditional fibre sensors. The main index-sensing mechanism of NCF is based on the wavelength shift of multimode signals' interference (MMI). When the incident light wave propagates along the axial direction of NCF, the outer medium of the adjacent substance will sense a lower index compared to the cladding layer index of the fibre. It facilitates a total internal reflection as a result of the Frensel reflection. In addition, the peak wavelength of the sensor can be changed by simply changing the length of the NCF.

FBG sensors can be incorporated into optical fibre cables. These sensors can be embedded into a new structure or surface-bonded onto an existing structure. This allows real-time monitoring of structure which will ultimately lead to truly smart sensor and provide fatigue data for subsequent analysis. As a result, FBG is able to play a crucial role in sensing technology. Nowadays, decent and sophisticated techniques in temperature measurement have become very important and necessary for safety precautions in industrial requirements.

1.2 Optical Fibre Bragg Grating and No-Core Fibre Sensors

Among the main motivations or objectives of writing this book is to share knowledge on designation, construction, and development of temperature sensor system using FBG and NCF sensors in indoor- and outdoor-based environments. Until now, a commercial germanium-doped silica fibre FBG is used and incorporated with a multimode NCF. This book will discuss the performance and uses of FBG and NCF in daily routines. The effect of temperature variations on the characteristics of FBG and NCF temperature sensors will be discussed and explained in detail. How do the changes of Bragg wavelength, bandwidth, and reflectivity of FBG and NCF respond to temperature variation? How do the characteristics of FBG sensor respond to the changes in temperature under different environmental conditions and placement heights? How do the FBG and NCF react when pressure is applied to the sensors?

This book will discuss the design, construction, experiment, and measurements of the performance and sensitivity of FBG and NCF sensors in indoor and outdoor environments. This book will also discuss the performance of FBG and NCF sensor systems at different placement heights and pressure. This is important to evaluate the effect of location and pressure on the FBG and NCF for multiple purposes. In the end, the efficiency of the focusing elements on the FBG and NCF sensor heads, based on transmission and reflection spectra will be discussed. This book enables us to understand the performance of FBG and NCF sensors by examining the characteristics and properties of the spectra. Different types of liquids and sudden changes in the outdoor environment such as temperature variation, rain, and wind might affect the stability of the sensor. With the increasing need to monitor structures such as bridges, tunnels, highways, dams, aircraft wings, and spacecraft fuel tanks, it is imperative to design and develop an effective sensor system which can detect any sudden changes in strain, pressure, and temperature. For this reason, the knowledge and evaluation of FBG and NCF sensors are very crucial. This book will deepen and enhance understanding of FBG and NCF temperature-sensing performance. The effect of the properties of FBG and NCF in three different liquids and temperatures will be examined. This enables us to apply the FBG and NCF temperature sensors in various fields, such as medicine, construction, manufacturing, industry, and many more.

1.3 Organization of Book

This book is divided into six chapters, beginning with a brief introduction and the overview of FBG and NCF in this Chapter. The review of the fundamentals of FBG and NCF and the overview of FBG and NCF for temperature sensing are provided in Chap. 2. In Chap. 3, the theoretical discussion of FBG and NCF will take place. This includes the coupled-mode theory (CMT) analysis, properties, optical response, and characteristics of FBG and NCF. Chapter 4 explains the experimental setup and methodology used for the investigations on FBG and NCF. The results, analysis, and discussion of the performance evaluation of FBG and NCF sensing in the indoor and outdoor environment will be discussed in Chap. 5. The Bragg and no-core wavelength response with respect to the temperature variation will be analyzed. Finally, the book is concluded in Chap. 6.

Chapter 2
Operational Principles of Fibre Bragg Grating and No-Core Fibre

2.1 Fibre Bragg Grating

Fibre optics is an overlap of applied science and engineering concerned with the design and application of optical fibre. Normal optical fibre s possess a uniform refractive index along their lengths. Such fibre is referred to as fibre Bragg grating (FBG). FBG is defined as a periodic perturbation of a refractive index formed by exposing its naked core to an intense optical interference pattern. FBG is a passive optical component which selectively reflects and transmits lights at certain wavelengths. The portion of light where the wavelength is equal to the Bragg wavelength will be reflected and the rest will be transmitted through the FBG. The refractive index variation scatters light that passes through the fibre. It provides modulation of core refractive index for single-mode fibre. In FBG, the gratings are uniformly spaced regions in fibre where the refractive index has been raised from the rest of the core. These radiations scatter light and they are called the Bragg effect. Every time the light hits the region of the scattered higher refractive index, a few light beams will be scattered from each higher index zone, interfering constructively and producing strong refraction as shown in Fig. 2.1. High-index regions scatter light at other wavelengths, but the scattered wave differs in phase by canceling each other out in destructive interference.

These nonresonant wavelengths are transmitted through the grating with low losses along the fibre. It is one of the most important examples of optical fibre designed to serve the function of optional components and emerges as an important technology in fibre optic sensing. Gratings in FBG are induced by exposing the fibre core to a periodic pattern of UV light over an extended time and the prolonged exposure results in FBG, as shown in Fig. 2.2.

© The Author(s) 2018
S. Daud and J. Ali, *Fibre Bragg Grating and No-Core Fibre Sensors*,
SpringerBriefs in Physics, https://doi.org/10.1007/978-3-319-90463-4_2

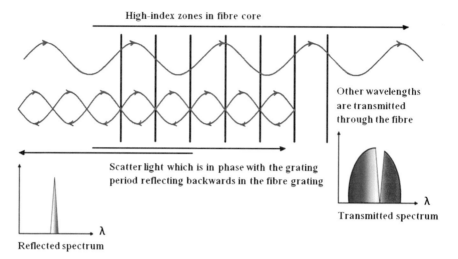

Fig. 2.1 Reflected and transmitted light in fibre Bragg grating

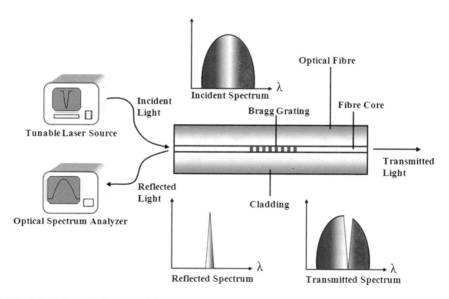

Fig. 2.2 Schematic diagram of fibre Bragg grating system

2.2 Writing Technique

There are a few ways to fabricate the Bragg grating, including internal and external writing technique, sequential writing technique, photomask technique, and point-by-point technique. The first in- fibre Bragg grating was demonstrated by

Ken Hill in 1978. Initially, the gratings were fabricated using a visible laser propagating along the fibre core. In 1989, Gerald Meltz and his colleagues demonstrated a considerably more flexible transverse holographic inscription technique where the laser illumination came from the side of the fibre. This technique uses the interference pattern of ultraviolet laser light to create the periodic structure of the fibre Bragg grating. This section will discuss the internal and external writing techniques of FBG, the easiest and most economical technique.

2.2.1 Internal Writing Technique

Bragg gratings were first fabricated using internal writing technique. Figure 2.3 shows the schematic diagram used for generating self-induced Bragg grating in internal writing technique. Intense argon-ion laser radiations with 488 nm wavelength are launched into a short piece of germanium-doped fibre for a few minutes. An increase in reflected light intensity occurs and continues to grow until almost all the lights are reflected from the fibre. The coherent light propagating in the fibre together with a small amount of light are reflected back from the end of the fibre which generates a standing wave pattern in the fibre. This standing wave pattern creates an index modulation in the fibre core through photosensitivity method.

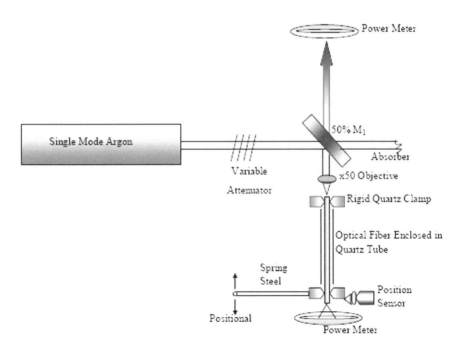

Fig. 2.3 Schematic diagram used for generating self-induced Bragg grating

Special measurements confirm that a very narrow band Bragg grating filter is formed with 90% reflectivity over the entire one-meter length of the fibre. It is shown that this grating can be used as a feedback mirror for a laser or as a sensor for strain measurements by stretching the fibre. Unfortunately, it reflects light only at the wavelength of the writing light. This limitation is then overcome ten years later after the external writing technique was invented by Hill and his co-workers. Internal writing technique is not commonly used due to the inefficiency of the writing process. This is because the Bragg grating is formed by internal writing. The limitation is the wavelength of the reflected light too close to the wavelength at which they are written.

2.2.2 External Writing Technique

The external writing technique was first reported in 1988. External writing technique can be classified into three groups, which are interferometric technique, point-by-point technique, and phase-mask technique. The techniques differ in the principle of writing and consequently, the equipment used in the writing process. External writing technique is more used compared to internal writing technique due to its ability to overcome the limitation of internal writing technique. For this reason, the transverse holographic technique is used. In this technique, fibre is exposed externally from the side to an interference pattern formed by two coherent UV beams. The period of photo-induced grating depends on the angle between the two interfering UV beams. In principle, Bragg gratings can be made to function at arbitrary wavelengths by adjusting the angle of the interference between the UV beams. This basic concept of FBG fabrication makes them suitable for various applications.

An alternative setup based on a prism interferometer is shown in Fig. 2.4. In this system, UV beam is directed through the hypotenuse face of a right-angle prism towards the apex in such a way only half of the beam is incident on the bank face. The other half of it undergoes reflection from the side face. The reflection of the beam is directed towards the back of prism where it can interfere with rest of the beam. In this case, the interference pattern is generated using total internal reflection from the prism that might be in contact with the fibre. The interferometer is very stable and allows the fibre to be illuminated for a long time. However, both the prism interferometer and the transverse holographic techniques require careful angular alignment of the interferometer to monitor the Bragg wavelength.

2.3 Historical Perspective of Fibre Bragg Grating Sensor

FBG is a type of fibre optic sensing which can work as a supersensitive sensor due to the extra-sensitive gratings in the optical fibre itself. FBG works as a sensor when changes in a particular environmental variable are correlated with Bragg

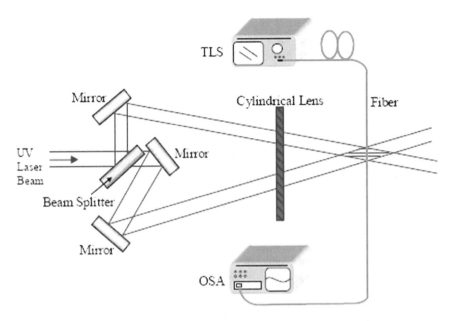

Fig. 2.4 Schematic diagram of the setup used for transverse holographic technique

wavelength shifts in the reflection and transmission wavelengths of FBG, such as refractive index, strain, and temperature. With the existence of Bragg grating in an optical fibre, the grating reflects and transmits narrow bandwidth of light centered at the Bragg wavelength within the transmission spectrum. The reflection and transmission wavelengths are dependent on the Bragg grating period and the guiding properties of the fibre.

Representative values of the strain- and temperature-induced wavelength shift for the grating produced in silica fibre are 1 and 10 pm $°C^{-1}$ at 1300 nm wavelength respectively. The use of such devices for temperature sensor has been reported in the 1990s by Jung and team. These results and observations were in good agreement with the research work on FBG temperature sensing done by Edmon and team in 2004. Selected FBG temperature sensor with the sensitivity of 16.5 ± 0.1 pm $°C^{-1}$ and in the range of −60 °C to 150°C at 1550.63 nm wavelength has been designed and developed by Ramesh and his team in 2006. In 1981, Lam and Garside successfully proved the relationship between photo-induced refractive index and strength of UV light exposure. This led to the discovery of a new technique that led to the external fabrication of FBG in germanium-doped silica fibre.

The side-writing technique was discovered by Meltz and his team in 1989, and greatly simplified the previous Hill's method. Since Meltz's work was published, new technologies for fabricating and producing FBGs have been developed rapidly. The new techniques overcome the complexity of FBG manufacturing process and make them reproducible at lower costs and shorter time. Finally, FBG becomes beneficial in many commercial applications in optical fibre communications and sensor systems, including their potentials in temperature and strain sensing. Following the realization of low-loss optical waveguides in the 1960s, optical fibre fibres have been developed to the point where they are now synonymous with modern telecommunication and optical sensor networks. The success in optical fibre properties such as low transmission loss, high optical damage threshold, and low optical nonlinearity renders them useful for telecommunications and sensing technologies.

In 1999, Ramesh and his co-workers designed and fabricated an FBG temperature sensor with a wavelength of 1550 nm by using the phase-mask technique. Excellent linearity between the shift in the center wavelength and temperature were observed in temperatures ranging from −60 to 150 °C. The center wavelength of the FBG was observed at 1550.36 nm. Its performance was characterized by monitoring the shift in the Bragg wavelength as a function of temperature. This is described by a linear equation given by $T = KL + b$, where T represents the temperature in degree Celsius, K is the thermal expansion coefficient, L is the position of the Bragg wavelength measured using an optical spectrum analyzer, and b is the wavelength position that corresponds to 0 °C temperature. It exhibits a shift in the center wavelength of 0.01 nm °C^{-1} change in every single temperature change. This means, for every 1 °C temperature change, it leads to a center wavelength shift of 0.01 nm.

In 2004, Edmon and co-workers investigated the transverse load and temperature sensitivity of fabricated FBG in a range of commercially available stress and geometrically induced high birefringent (HiBi) fibre. The reflected wavelengths of the FBG in each Eigen mode polarization were measured independently and simultaneously using a costume-designed interrogation system. HiBi FBG for temperature sensor with a maximum sensitivity of 11.5 ± 0.1 pm °C^{-1} was investigated. The temperature sensitivity of 9.0 pm °C^{-1} at a low temperature which changed to 17.5 pm °C^{-1} at 1000 °C was reported in 2004. The highest sensitivity of 16.5 pm °C^{-1} was measured using the HiBi FBG fabricated in a PANDA ring resonator fibre. This value was approximately 27% greater than any other types of fibre. The physical geometry of the fibre core and cladding, and the presence of stress-applying parts, influenced the sensitivity of FBG.

In 2005, Hilaire and team wrote an FBG at an arbitrary wavelength without extensive recalibration and reconfiguration of the writing equipment in some embodiments. A pair of writing beams was used to expose the fibre and the crossing angles of the writing beams can be adjusted. In the same year, a group of researchers at NASA Glenn Research Center evaluated the performance of a device at high temperature up to 1000 °C using commercially available polyimide-coated high-temperature gratings to assess the stability of FBG at high temperature. The gratings were placed in a furnace and the generated signals were sent to a photodetector, and from there, to a spectrum analyzer. The signals generated were then fed into a computer equipped with LabVIEW software. The software was used to control and monitor the equipment, as well as to process the data taken. Tests including thermal cycling from room temperature to 750 °C and up to 1000 °C, as well as prolonged exposure of the gratings to 1000 °C were completed. The tests confirmed the formation of secondary thermally stable gratings in the germania-doped glasses at high temperature. The secondary gratings were formed in place of the primary ones originally written by the UV light. The secondary gratings became dominant at high temperatures.

In 2007, Bowei and Mojtaba successfully demonstrated that molecular water FBG possessed a higher sensitivity compared to that of a high temperature-resistant FBG fabricated using hydrogen-loaded conventional FBG which has been given thermal annealing treatment at a temperature around 1000 °C. The shift in Bragg wavelength with the change of temperature was used as an effective factor for the temperature sensing on temperature compensation in other sensing, including the ones in their research. The sensitivities of FBGs were determined by observing the shift of Bragg wavelength with temperature change. The variations of Bragg wavelength shift with respect to the temperature change were almost linear at higher temperature ranges. The sensitivity of the sensor increased from 9.0 pm $°C^{-1}$ at low temperature and up to 17.5 pm $°C^{-1}$ at temperatures around 1000 °C, whereas the conventional hydrogen-loaded FBG possessed 16.6 pm $°C^{-1}$ as compared with 15.0 pm $°C^{-1}$ at temperatures around 700 °C.

In 2008, Ho fabricated a single-mode FBG using phase-mask technique. KrF excimer laser was used as the UV source. The FBGs were fabricated with different pulse energies and exposure times. The sensitivity of FBG was determined using three different mediums, namely liquid, air, and metal. From the research done, it was concluded that the phase-mask technique is suitable for FBG fabrication and FBG can be used as a sensor in different media. This research was then continued by Daud in 2010. He investigated the sensitivity of commercial and fabricated FBG in harsh environments, including in high temperature and rainy and windy weather. It was found that FBG can be used in a variety of environments.

2.4 Why Fibre Bragg Grating Sensor?

The existence of FBG in the fields of sensor technologies is very much appreciated. FBG reflects a narrow spectral part of light guided in the fibre core at the Bragg wavelength, which is dependent on the grating period and the refractive index of fibre core itself. The grating period is highly sensitive to small changes in mechanical strain, temperature, humidity, and other physical parameters, including the refractive index of the fibre. What distinguishes the FBG sensor from conventional electrical strain gauges are its excellent linear characteristics, immunity to electromagnetic interference, low fibre loss (over many kilometers), highly multiplex distributed sensor arrays, and intrinsic safety in explosive environments. FBG sensor gives high accuracy, sensitivity, and immunity to radio-frequency interference. It has the ability to be made into a compact, lightweight, and rugged device small enough to be embedded and laminated into structures or substances to create smart materials that can operate in harsh environments, such as underwater, where conventional sensors cannot work.

FBG sensor has the ability to accommodate multiplexing and an inherent low transmission loss at 1550 nm wavelength, which allows one to use many sensors on a single optical fibre at arbitrary spacing. In other words, it can be highly multiplexed; the installation and use of FBG are very straightforward. With the grating multiplex on a single fibre, one can access many sensors with a single connection to the optical source and detector. It gives the potential low cost and results of the high-volume automated manufacturing process. FBG can also be used in explosive atmospheres, either in natural gas or oil. FBG has rugged passive components, resulting in a high lifetime. It is small in size and can be easily embedded into the desired areas. It forms an intrinsic part of the fibre optic cable that can easily transmit measurement signal over several kilometers. There are no interferences with electromagnetic radiation, so it might work in many hostile environments where conventional sensors would fail. FBG does not use electrical signals as well, thus making it intrinsic safety. It has the ability to multiplex many sensors using only one optical fibre, driving down the cost of the complex control system. FBG is highly reliable and cost-effective. It can produce fast and accurate measurements and does not require on-site power supply. It is highly sensitive, thus having a faster response time.

The criteria of fibre optic sensors include wide bandwidth, compactness, geometric versatility, and economic. It is characterized by high sensitivity compared to other type of sensors. It is passive in nature due to the dielectric construction. Specially prepared fibre can withstand high temperature and harsh environments. In telemetry and remote sensing applications, it is possible to use a segment of the fibre as the sensor gauge while a long length of fibre can convey the sensed information to a remote station. Signal processing devices like splitters, combiners, multiplexers, filters, and delay lines are mostly made of fibre elements, thus enabling the realization of an all—fibre measuring system. Recently, photonic circuits have been proposed as a single-chip optical device or signal processing elements which enable miniaturization, batch production, economic, and enhanced properties. Optical fibre, being a physical medium, is subjected to perturbation of a kind or other at all times, therefore experiences geometrical (size, shape) and optical (refractive index, mode conversion) changes to a larger or lesser extent depending on the nature and magnitude of the perturbation. In communication applications, one tries to minimize such effects so that signal transmission and reception are reliable. On the other hand, the response to external influence is deliberately enhanced so that the resulting change in optical radiation can be used as a measure of the external perturbation in fibre optic sensing systems. In communication systems, the signal passing through a fibre is already modulated. In sensing systems, the fibre acts as a modulator. It serves as a transducer and converts measurements like temperature, stress, strain, rotation, and magnetic currents into a corresponding change in the optical radiation. Since light is characterized by amplitude, phase, frequency, and polarization, any one or more of these parameters may undergo a change. Fibre optic sensors have wide bandwidth, compactness, geometric versatility, and economic advantage. In general, fibre optic sensors are characterized by high sensitivity compared to other types of sensors. In telemetry and remote sensing applications, it is possible to use a segment of the fibre as the sensor gauge while a long length of the same or another fibre can convey the sensed information to a remote station. Development of distributed and array sensors covering extensive structures and geographical locations is also feasible. Many signal processing devices like splitter, combiner, multiplexer, filter, and delay line can also be made from fibre elements thus enabling the realization of an all—fibre measuring system. Recently, photonic circuits have been proposed as a single-chip optical device or signal processing element which enables miniaturization, batch production, economical, and has enhanced capabilities.

2.5 No-Core Fibre Sensor

No-core fibre (NCF) is a highly pure silica fibre without cladding. The main index-sensing mechanism of NCF is based on the wavelength shift of multimode signals' interference (MMI). When the incident light wave propagates along the axial direction of NCF, the outer medium of the adjacent substance is sensed with a lower index than the cladding layer index of the fibre and facilitates a total internal reflection due to Fresnel reflection. Since the structure of the NCF comes with no cladding, NCF exhibits a more sensitive response to the changes of the environment. The jointing point with single-mode fibre (SMF) is a taper, which is easier for light to couple into the NCF and will improve the sensing sensitivity. The peak wavelength of the sensor can be changed by simply changing the length of the NCF. Figure 2.5 shows the schematic diagram of NCF where the multimode fibre (MMF) is jointed with SMF at both ends. When light is injected from the input SMF into the NCF, multiple modes will be excited and will propagate along the NCF. The theoretical peak wavelength indicates response to the NCF sensor. The NCF sensor is very sensitive to any small changes in the length of NCF.

Fig. 2.5 No-core fibre

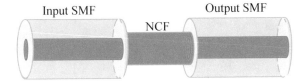

Chapter 3
Theory

3.1 Coupled-Mode Theory

One of the methods commonly used to analyze and investigate light propagation in perturbed coupled waveguides is coupled-mode theory (CMT) method. The basic idea of CMT method is that the modes of unperturbed (or uncoupled) structures are defined and solved first. Linear combinations of these modes are used as trial solutions to Maxwell's equations for the complicated perturbed or coupled structure. This theory assumes the field of modes of unperturbed structures. In many practical cases, this assumption is valid and does give an insightful and often accurate mathematical description of electromagnetic wave propagation. CMT method is often used as a technique to obtain quantitative information on the diffraction efficiency and spectral dependence of fibre gratings. Due to its simplicity and accuracy in modeling and the optical properties of most fibre gratings, it is one of the most popular techniques utilized in describing the behavior of Bragg gratings.

Commonly, the term mode coupling addresses one of at least three different means of power transfer

i. Coupling modes of distinct waveguides by evanescent fields.
ii. Coupling modes in the same waveguide by longitudinally homogeneous perturbation.
iii. Co- and contra-directional coupling by longitudinal inhomogeneous, usually short periodic perturbations.

Fibre Bragg Grating (FBG) is based on contra-directional couplings. In the case of single-mode fibre, the propagating core mode is reflected into the identical core mode propagating in the opposite direction. In most cases of moderated FBGs, the coupling of Bragg reflection is dominant compared to the cladding-mode couplings, even though it is possible for the core mode to be coupled with the counterpropagating cladding modes in cases that include strong gratings and blazed grating. The schematic illustration of contra-directional coupling in optical fibres is shown in Fig. 3.1.

© The Author(s) 2018
S. Daud and J. Ali, *Fibre Bragg Grating and No-Core Fibre Sensors*,
SpringerBriefs in Physics, https://doi.org/10.1007/978-3-319-90463-4_3

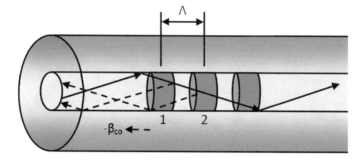

Fig. 3.1 Illustration of contra-directional coupling

Based on contra-directional coupling, the nominal Bragg wavelength is given as

$$\lambda_{B,0} = 2n_{eff}\Lambda \tag{3.1}$$

and

$$\lambda_{max} = 2\left(n_{eff} + \delta n_{eff,av}\right)\Lambda = \left[1 + \frac{\delta n_{eff,av}}{n_{eff}}\right]\lambda_{B,0}, \tag{3.2}$$

where n_{eff} is the unperturbed effective index of the core and Λ is the grating period.

3.2   Principle Operation of Fibre Bragg Grating

The basic principle of FBG-based sensor lies in the wavelength shift of the perturbed Bragg signal as a function of measuring elements such as strain, temperature, and force. It is related to the refractive index of the material and grating pitch itself. Sensor systems involves such gratings which usually work by injecting light from a spectral component at the Bragg wavelength. The strong reflected light from an incident broadband source over a narrow wavelength range and transmitted without changing all other wavelengths, produce a dip in the transmission and reflection spectra at the same wavelength range. One must first consider an unstrained low birefringence single-mode fibre for which the refractive index of the cross-section of the core is uniform. In principle, the reflection spectrum of FBG is complimentary to its transmission spectrum, although this is not realized in the practice of transmission losses, which may result from the conversion of the coupling to cladding modes.

The existence of Bragg grating in an optical fibre may reflect a very narrow bandwidth of light centered at the Bragg wavelength within the transmission spectrum. The reflected wavelength depends on the period of Bragg grating and guiding properties of the fibre. Both quantities are proportionally related to environmental variables such as temperature and strain variation. FBG operates as a sensor when changes in a particular environmental variable are correlated with a shift in the reflected wavelength of the FBG. When a guided fibre mode at the Bragg condition is incident upon an FBG, a certain percentage of incident light will be scattered at each grating plane. For certain directions, the wavelets are created and each plane is in phase.

Figure 3.2 shows the schematic diagram of an enlarged FBG. As indicated in the figure, λ_{in} is the wavelength of incident light, $\lambda_B = 2\Lambda_g n_{eff}$ is the Bragg wavelength or center wavelength of the FBG, Λ_g is the spacing of grating, δn_{eff} is the change of effective refractive index of the fibre, n_{eff} is the effective refractive index of fibre core at free-space center wavelength, k_1 is the modal wavevector of the forward-propagating wave, k_2 is the modal wavevector of the backward-propagating wave, and the grating momentum, $K = 2\pi/\Lambda_g$ for the FBG.

3.2.1 Bragg Wavelength

The interaction of the light traveling through the FBG depends on its wavelength. Bragg wavelength may be altered with changes in the applied strain, ambient temperature, and incident wavelength. With the appearance of the gratings in the photosensitive fibre during irradiation, the initial Bragg wavelength is measured in the last exposure laser. The Bragg wavelength shift is the difference between two measured Bragg wavelengths. It is given as

$$\Delta\lambda_B = \lambda_{in} - \lambda_B \qquad (3.3)$$

The change in Bragg wavelength of FBG due to temperature can be measured using the relation

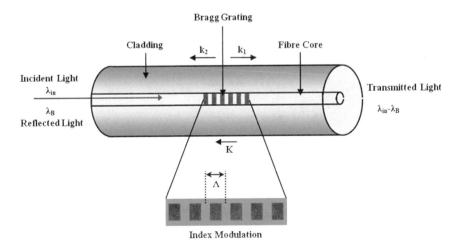

Fig. 3.2 Schematic diagram of enlarged FBG

$$\Delta\lambda_{\mathrm{B}} = \lambda_{\mathrm{B}}(\xi + \alpha)\Delta T \tag{3.4}$$

where $\Delta\lambda_{\mathrm{B}}$ is the Bragg wavelength shift, λ_{B} is the Bragg wavelength of the FBG, ξ is the thermo-optic coefficient, α is the thermo-expansion coefficient, and ΔT is the change of temperature. For example, the thermo-optic and thermo-expansion coefficients of germanium-doped silica core optical fibre are given as $\xi = 8.6 \times 10^{-6}\,°\mathrm{C}^{-1}$ and $\alpha = 0.55 \times 10^{-6}\,°\mathrm{C}^{-1}$ respectively.

When a guided mode at the Bragg condition is incident upon an FBG, a certain percentage of the incident light will be scattered at each grating plane. Certain directions may be discovered where wavelets created at each plane are in phase. If these directions correspond to a mode of the fibre, then a resonant condition is satisfied. In this case, the forward wave momentum, $\vec{k}$ is reflected by the grating momentum, $\vec{K}$.

The conservation of momentum can be dictated as

$$\vec{k_2} = \vec{k_1} + \vec{K}, \tag{3.5}$$

where $\vec{k_1}$ is the modal wavevector of the forward-propagating wave, $\vec{k_2}$ is the modal wavevector of the backward-propagating wave, and $\vec{K}$ is the grating momentum, where $\vec{K} = \frac{2\pi}{\Lambda}$.

As the photon frequencies are identical for the two propagating waves, $\lambda_1 = \lambda_2 = \lambda_{\mathrm{B}}$, therefore, from Eq. (3.5) we can get

$$2\left(\frac{2\pi n}{\lambda_{\mathrm{B}}}\right) = \frac{2\pi}{\lambda} \tag{3.6}$$

which simplifies to the first-order Bragg condition as

$$\lambda_{\mathrm{B}} = 2n_{\mathrm{eff}}\Lambda, \tag{3.7}$$

where the Bragg wavelength, λ_{B}, is the free-space center wavelength of the input light that will be back-reflected from the Bragg gratings, and n_{eff} is the effective refractive index of the fibre core at the free-space center wavelength.

The effective refractive index of the fibre core, n_{eff}, is given as

$$n_{\mathrm{eff}}(z) = \Delta n \, \cos\left(\frac{2\pi z}{\Lambda}\right) \tag{3.8}$$

where Δn is the amplitude of index modulation.

3.2.2   Bandwidth of Bragg Grating

In 2002, Hill defined another FBG property: its bandwidth, which is a measure of the wavelength range in which the grating reflects light. Previously in 1999, Grattan and Meggit defined bandwidth as the separation in the wavelength between two points on either side of Bragg wavelength where the reflectivity has decreased to half of its maximum value. The example of bandwidth measurement of a fibre is shown in Fig. 3.3.

The full width at half maximum, $\Delta\lambda_{\text{FWHM}}$, of the central reflection peak is the easiest way to measure the bandwidth of fibre grating. FWHM is defined as the wavelength interval between the 3-dB points of fibre. However, it is more convenient to determine the bandwidth of the grating by using

$$\Delta\lambda_{\text{o}} = \lambda_{\text{o}} - \lambda_{\text{B}} \tag{3.9}$$

where λ_{o} is the wavelength of the first zero in reflection spectra and λ_{B} is the Bragg wavelength.

If the grating is a weak grating or induces a small index change, then the reflectivity can be simplified using Born approximation. It is given as

$$R(L, \lambda) = (\Delta n)^2 \frac{\sin^2(\Delta kL)}{(\Delta kL)^2} \tag{3.10}$$

From this expression, it may readily be seen that the FWHM of the peak between the first zero occurs when

$$\Delta kL = \pm\pi \tag{3.11}$$

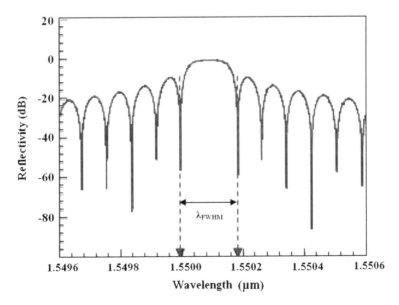

Fig. 3.3 Calculation of FWHM using reflection spectrum

By substituting Eq. (3.11) into Eq. (3.10), it can be rewritten as

$$\frac{2\pi n_{\mathrm{co}}\Delta\lambda_B}{\lambda_B^2}L = \pm\pi \tag{3.12}$$

Thus, the full spectral width of the Bragg grating is given as

$$\Delta\lambda_B = \frac{\lambda_B^2}{n_{\mathrm{co}}L} \tag{3.13}$$

In the frequency domain, it can be expressed as

$$\Delta v_B = \frac{c}{n_oL} \tag{3.14}$$

Although the above expressions are derived with the initial assumption that the grating is weak, the important trend to note is that the grating bandwidth decreases with increasing grating length. It can be seen that the bandwidth of Bragg grating depends upon two factors, namely, the number of grating planes, N and the depth of index modulation, Δn.

In 1999, Kashyap found a more accurate expression for full width at half-maximum bandwidth, $\Delta\lambda_{\mathrm{FWHM}}$ of a grating. It is given as

$$\Delta\lambda = \lambda_B s\sqrt{\left(\frac{\Delta n}{2n_o}\right)^2 + \left(\frac{1}{N}\right)^2} \tag{3.15}$$

where λ_B is the Bragg wavelength, N is the number of grating plane, n_o is the core index which the typical value of 1.45, Δn is the depth of index modulation, and $s \sim 1$ for strong gratings (with more than 100 gratings), and $s \sim 0.5$ for week gratings.

3.2.3 Reflectivity of Bragg Grating

As discussed earlier in the first chapter, FBGs are based on the principle of Bragg reflection. When the light propagates periodically alternating regions of higher and lower refractive index, it is partially reflected at each interface between those regions. If the space between those regions are such that all the partial reflections add up in phase, the total reflection can grow up to nearly 100% even though the individual reflections are very small. This condition will only take place at the specific wavelength. For the other wavelengths, the out-of-phase reflections end up canceling each other, resulting in high transmission.

Consider a uniform Bragg grating formed within the core of an optical fibre with an average refractive index, n_o. The refractive index profile can be expressed as

$$n(z) = n_{\mathrm{o}} + \Delta n \cos \frac{2\pi z}{\Lambda} \qquad (3.16)$$

where Δn is the amplitude of induced refractive index perturbation and z is the distance along the fibre longitudinal axis. The value of Δn is in range of 10^{-5}–10^{-3}.

Based on CMT, for constant modulation amplitude and period, the reflectivity of a grating is given as

$$R(L, \lambda) = \frac{\Omega^2 \sin g^2(sL)}{\Delta k^2 \sinh^2(sL) + s^2 \cosh^2(sL)} \qquad (3.17)$$

where $R(L, \lambda)$ is the reflectivity (in the function of the grating length, L, and wavelength, λ), Ω is the coupling coefficient, $\Delta k = k - \pi/\lambda$ is the detuning wavevector, $k = (2\pi n_{\mathrm{o}})/\lambda$ is the propagation constant, and $s^2 = \Omega^2 - \Delta k^2$.

The coupling coefficient, Ω, for the sinusoidal variation of index perturbation along the fibre axis is given as

$$\Omega = \frac{\pi \Delta n}{\lambda} M_{\mathrm{p}} \qquad (3.18)$$

where M_{p} is the fraction of the fibre mode power contained by fibre core.

There is no wavevector detuning at the center wavelength of Bragg grating, and $\Delta k = 0$. Therefore, from Eq. (3.17), the expression for the reflectivity of FBG becomes

$$R(L, \lambda) = \tanh^2(\Omega L) \qquad (3.19)$$

The reflectivity of FBG increases as the induced index of refraction increases. Similarly, as the length of grating increases, the reflectivity will also increase. The reflectivity level increases as the induced index of refraction change increases. Similarly, as the length of the grating increases, so does the resultant reflectivity. The side lobes of the resonance are due to multiple reflections to and from opposite ends of the grating region.

The general expression for the approximate full width at half-maximum bandwidth (λ_{FWHM}) of a grating is given as

$$\lambda_{\mathrm{FWHM}} = \lambda_{\mathrm{B}} s \sqrt{\left(\frac{\Delta n}{2n_{\mathrm{o}}}\right)^2 + \left(\frac{1}{N}\right)^2}, \qquad (3.20)$$

where N is the number of the grating planes. The parameter s is ~ 1 for strong gratings (for gratings with near 100% reflection) and ~ 0 for weak gratings.

The maximum reflectivity of FBG can be calculated by measuring the transmission power dip as

$$R = \left(1 - 10^{-\frac{d}{10}}\right) \times 100, \tag{3.21}$$

where R is the reflectivity of FBG and d is the power dip.

The variation of the reflectivity with the wavelength depends on the grating. Fine, thin, evenly spaced lines tend to concentrate reflection at a narrow range of wavelengths. Turning up exposures to make a stronger grating will increase the reflectivity and broaden the range of reflected wavelengths. Commercial devices using this design select a range of wavelengths as narrow as a few tenths of nanometer ranging up to several nanometers wide. The narrow ranges are well matched to the requirements of wavelength division multiplexing in the 1550 nm band.

3.3   Optical Response of Bragg Grating

Since the discovery of photosensitivity in optical fibres and the UV side-writing technique, FBG has become an interesting area in the field of sensors around the world. The development effort has been principally led by applications to sense different measurements including temperature, strain, pressure, electric field, magnetic field, chemical concentration, and many more. These applications are based on the same principle, namely the measurement of Bragg wavelength shift of reflected maxima caused by the measurements.

By expanding Eq. (3.1), that is, $\lambda_{B,0} = 2n_{eff}\Lambda$, in terms of partial derivatives with respect to the variable length, temperature, and wavelength, the shift in Bragg grating wavelength can be derived as

$$\Delta\lambda_B = 2\left(\Lambda\frac{\delta n_{eff}}{\delta L} + n_{eff}\frac{\delta\Lambda}{\delta L}\right)\Delta L + 2\left(\Lambda\frac{\delta n_{eff}}{\delta T} + n_{eff}\frac{\delta\Lambda}{\delta T}\right)\Delta T + 2\left(\Lambda\frac{\delta n_{eff}}{\delta\lambda} + n_{eff}\frac{\delta\Lambda}{\delta\lambda}\right)\Delta\lambda \tag{3.22}$$

3.3.1   Optical Response to Wavelength

From Eq. (3.22), the optical response of fibre grating to the shift in wavelength is written as

$$\Delta\lambda_B = 2\left(\Lambda\frac{\delta n_{eff}}{\delta\lambda} + n_{eff}\frac{\delta\Lambda}{\delta\lambda}\right)\Delta\lambda \tag{3.23}$$

where $\Delta\lambda$ is the change in wavelength of the grating due to initial wavelength.

In reality, the variation of refractive index due to the change in wavelength is negligible. In addition, the periodic spacing of the index modulations in fibre is unaffected by the wavelength change. Thus, by neglecting the wavelength effects, Eq. (3.22) can be written as

$$\Delta\lambda_B = 2\left(\Lambda\frac{\delta n_{\text{eff}}}{\delta L} + n_{\text{eff}}\frac{\delta\Lambda}{\delta L}\right)\Delta L + 2\left(\Lambda\frac{\delta n_{\text{eff}}}{\delta T} + n_{\text{eff}}\frac{\delta\Lambda}{\delta T}\right)\Delta T \tag{3.24}$$

where ΔL is the change of length and ΔT is the change of temperature of the fibre.

3.3.2 Optical Response to Strain

The effects of the strain on Bragg grating are twofold. First, the change in physical spacing between successive index modulations will cause a shift in Bragg wavelength. Second, the strain-optic effect will induce a change in refractive index, thus causing a shift in Bragg wavelength.

The change in the center wavelength of Bragg grating for a given change in strain is given as

$$\Delta\lambda_B = 2\left(\Lambda\frac{\delta n_{\text{eff}}}{\delta L} + n_{\text{eff}}\frac{\delta\Lambda}{\delta L}\right)\Delta L \tag{3.25}$$

where ΔL is the change in physical length of the grating due to the strain applied.

Given that the path integrated by longitudinal strain is given by $\varepsilon_z = \Delta L/L$, Eq. (3.25) can be rewritten as

$$\Delta\lambda_B = 2\Lambda\frac{\delta n_{\text{eff}}}{\delta L}\Delta L + 2n_{\text{eff}}\frac{\delta\Lambda}{\delta L}\varepsilon_z L \tag{3.26}$$

Defining,

$$\Delta n_{\text{eff}} = \frac{\delta n_{\text{eff}}}{\delta L}L \tag{3.27}$$

and

$$\Delta\left(\frac{1}{n_{\text{eff}}^2}\right) = -\frac{2\Delta n_{\text{eff}}}{n_{\text{eff}}^3} \tag{3.28}$$

By substituting Eqs. (3.27) and (3.28), Eq. (3.26) can be written as

$$\Delta\lambda_B = 2\Lambda\left[-\frac{n_{\text{eff}}^3}{2}\cdot\Delta\left(\frac{1}{n_{\text{eff}}^2}\right)\right] + 2n_{\text{eff}}\varepsilon_z L\frac{\delta\Lambda}{\delta L} \tag{3.29}$$

The shift in the center wavelength for a grating with a given center wavelength, λ_B, is given as

$$\frac{\Delta \lambda_B}{\lambda_B} = (1 - p_e)\varepsilon_z \qquad (3.30)$$

where ε_z is the strain applied and p_e is 0.213 for germanosilicate optical fibre.

Equation (3.30) can be simplified and rearranged as

$$\Delta \lambda_B = \lambda_B(1 - p_e)\varepsilon_z \qquad (3.31)$$

3.3.3 Optical Response to Temperature

Thermal expansion or contraction changes the grating period and will affect the FBG optical responses as well. Simply said, the Bragg wavelength, λ_B, and the effective refractive index, n_{eff}, are temperature dependent (thermo-optic effect). Therefore, the change in the Bragg wavelength for a given change in temperature is given as

$$\Delta \lambda_B = 2\left(\Lambda \frac{\delta n_{eff}}{\delta T} + n_{eff} \frac{\delta \Lambda}{\delta T} \right) \Delta T \qquad (3.32)$$

where ΔL is the change in physical length of the grating due to the temperature applied.

Defining,

$$\frac{\delta n_{eff}}{\delta T} = \xi n_{eff} \qquad (3.33)$$

and

$$\frac{\delta \Lambda}{\delta T} = \alpha \Lambda \qquad (3.34)$$

Substituting Eqs. (3.33) and (3.34) into Eq. (3.32) leads to

$$\Delta \lambda_B = 2\left(\Lambda \xi n_{eff} + n_{eff} \alpha \Lambda \right) \Delta T = 2 n_{eff} \Lambda \xi \Delta T + 2 n_{eff} \Lambda \alpha \Delta T \qquad (3.35)$$

Substituting $2 n_{eff} \Lambda = \lambda_B$ into Eq. (3.35) gives

$$\Delta \lambda_B = \lambda_B \xi \Delta T + \lambda_B \alpha \Delta T \qquad (3.36)$$

Rearranging Eq. (3.36), the Bragg wavelength shift, $\Delta \lambda_B$, becomes

$$\Delta \lambda_B = \lambda_B(\xi + \alpha)\Delta T \qquad (3.37)$$

For germanium-doped silica core, the thermo-optic coefficient, ξ is $8.6 \times 10^{-6}\,^{\circ}\mathrm{C}^{-1}$ and the thermal-expansion coefficient, α is $0.55 \times 10^{-6}\,^{\circ}\mathrm{C}^{-1}$.

3.4 Characteristics of Fibre Bragg Grating

Figure 3.4 illustrates the basic principle of FBG. The characteristics of FBG will be examined in terms of Bragg conditions, index modulation, transmission and reflection spectra, effective refractive index, and grating period.

3.4.1 Bragg Condition

A certain percentage of the incident light passing through the FBG will be scattered at each of the FBG's grating planes. Certain directions may be observed where wavelets created at each plane are in phase. A resonant condition is satisfied if these directions correspond to the mode of the fibre. In this case, the forward-propagating wave momentum, k, is reflected by grating momentum, K.

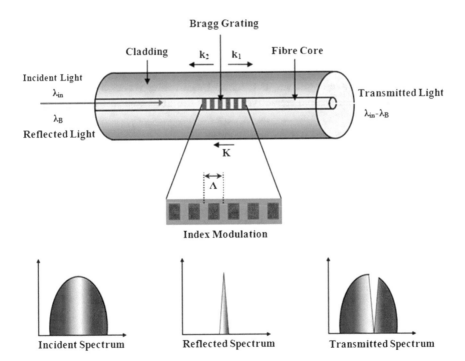

Fig. 3.4 Schematic diagram of fibre Bragg grating

The conservation of momentum indicates that

$$\vec{k}_2 = \vec{k}_1 + \vec{K},$$ (3.38)

where $\vec{k}_1$ is the modal wavevector of the forward-propagating wave, $\vec{k}_2$ is the modal wavevector of the backward-propagating wave, and grating momentum, $K = 2\pi/\Lambda$.

Both $\vec{k}_1$ and $\vec{k}_2$ are defined as

$$k_\alpha = \frac{2\pi}{\lambda_\alpha}$$ (3.39)

As the photon frequencies for two propagating waves are identical, the Bragg wavelength, λ_B is given as

$$\lambda_B = \lambda_1 = \lambda_2$$ (3.40)

Thus, from Eq. (3.38)

$$2\left(\frac{2\pi n}{\lambda_B}\right) = \frac{2\pi}{\lambda}$$ (3.41)

which can be simplified as the first-order of Bragg condition is defined as

$$\lambda_B = 2n_{\text{eff}}\Lambda,$$ (3.42)

where the Bragg wavelength, λ_B, is the free-space center wavelength of the input light and n_{eff} is the effective refractive index of the fibre core at the free space of center wavelength.

The effective refractive index, n_{eff}, is given as

$$n_{\text{eff}}(z) = n_{\text{co}} + \Delta n \cos\left(\frac{2\pi z}{\Lambda}\right)$$ (3.43)

where n_{co} is the core average index and Δn is the amplitude of index modulation of the fibre. For a given 1550 nm grating wavelength, the typical values of the aforementioned variables range from $n_{\text{co}} = 1.45$ to 1.50 and $\Delta n = 10^{-2}$ to 10^{-5}.

3.4.2   Transmission and Reflection in Fibre Bragg Grating

Each grating in FBG reflects a certain portion of the light at all wavelengths. If the wavelength passing through the glass is twice the spacing or period of the grating in fibre, all the scattered lights are in phase. The light waves interfere constructively and that wavelength is reflected. The light wave going through the region between the grating needs to be twice the distance between the gratings because one will go through the FBG and one will be reflected back. As the number of gratings increase, the spacing becomes more uniform and they are written more strongly, thus the reflection is stronger.

If Λ_g is the grating period and n is the refractive index of glass, the reflected wavelength in air is given as

$$\lambda_{\text{grating}} = 2n\Lambda_g \qquad (3.44)$$

Other wavelengths that do not meet this criterion are not reflected in phase. The scattered light waves do not add constructively. FBG can be made to have a peak reflection across a narrow band with nearly square sides as shown in Fig. 3.5. The transmission plot shows the transmission loss at the mentioned wavelength while the reflection curve shows the fraction of light reflected at that wavelength. The rest of light outside the selected band passes through unaffected.

The variation of FBG's reflectivity with the wavelength depends on the grating. Fine, thin, and evenly spaced intervals tend to concentrate reflection at a narrow range of wavelengths. Turning up exposures to make a strong grating will increase the reflectivity of FBG and broaden the range of reflected wavelength. The narrow range of wavelengths is well matched to the requirements of wavelength division multiplexing in 1550 nm band.

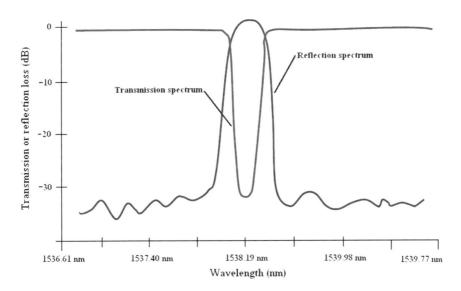

Fig. 3.5 Transmission and reflection in FBG

3.4.3   Effective Refractive Index

The effective refractive index, n_{eff}, is the ratio of the free-space velocity of propagation or guided velocity. It is different for each individual mode where generally the effective refractive index of FBG is defined as

$$n_{eff} = \frac{\lambda_B}{2\Lambda},\qquad(3.45)$$

where n_{eff} is the effective refractive index of FBG, λ_B is the Bragg wavelength, and Λ is the grating period of the fibre.

3.4.4   Refractive Index Modulation

If a region of the germanium-doped fibre core is exposed to a UV wavelength laser beam, the index of refraction can be changed in that region. By using UV light to expose the fibre from the side with a periodic pattern created by using an interferometer or phase mask, periodic modulation of the index of refraction can be generated along the fibre core. At each periodic index change, a small amount of the guided light is reflected as depicted in Fig. 3.6.

If the spacing of the index periods is equal to one half of the wavelength of the light, then the waves will interfere constructively (the round trip of each reflected wave is one wavelength) and a large reflection will occur from the periodic array. Optical signals whose wavelengths are not equal/different to one half of the spacing will travel through the periodic array unaffected. With a reasonable assumption that the gratings are all uniform, the amplitude of the refractive index modulation, Δn_{mod}, can be estimated using Eq. (3.46)

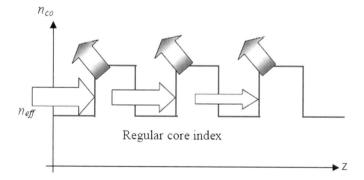

Fig. 3.6 Refractive index modulation of fibre

$$\Delta n_{\text{mod}} = \frac{\lambda}{n\eta L} \tanh^{-1}\left(\sqrt{R}\right) \tag{3.46}$$

When $\tanh^{-1}\left(\sqrt{R}\right) = \ln\left(\frac{1+\sqrt{R}}{1-\sqrt{R}}\right)$, based on the inverse hyperbolic, Eq. (3.46) can be rearranged and becomes

$$\Delta n_{\text{mod}} = \frac{\lambda}{2\pi\eta L} \ln\left(\frac{1+\sqrt{R}}{1-\sqrt{R}}\right), \tag{3.47}$$

where R is the reflectivity of uniform Bragg grating, λ is the grating operating wavelength, L is the length of the grating, and η is the mode overlap parameter (fraction of the fibre mode power contained in the core).

η is given by

$$\eta = \frac{\pi^2 d_{\text{co}}^2 \text{NA}^2}{\lambda^2 + \pi^2 d_{\text{co}}^2 \text{NA}^2}, \tag{3.48}$$

where d_{co} is the fibre core diameter and NA is the numerical aperture of the fibre.

In accordance with the phase-matching condition, a variation in the local effective refractive index, Δn_{eff}, will appear as a shift in the local Bragg wavelength, λ_{B}, such that

$$\Delta n_{\text{eff}} = \frac{\Delta \lambda_{\text{B}}}{2\Lambda} \tag{3.49}$$

where the Bragg wavelength shift during the grating fabrication, $\Delta \lambda_{\text{B}}$, is measured with the Bragg wavelength at the start of the inscription, λ_{bi} as a reference, and Λ is the grating period, i.e., half of the phase-mask period.

3.4.5 Grating Period

Grating period, Λ, is only decided by the phase-mask period, where the strength of grating is directly related to the efficiency of the beam which diffracted by phase mask in ±1 order. Grating period is directly proportional to the phase-mask period.

Grating period, Λ, is defined as

$$\Lambda = \frac{\Lambda_{\text{pm}}}{2}, \tag{3.50}$$

where Λ_{pm} is the phase-mask period.

3.5   Photosensitivity in Optical Fibre

Photosensitive fibres are one of the methods used to fabricate FBG. The formation of a permanent grating in an optical fibre was first demonstrated in 1978. It was shown that when intense UV light is radiated on an optical fibre, the refractive index of the fibre core changed permanently. This effect is termed as photosensitivity. It is known that germanium-doped silica fibre exhibits excellent photosensitivity. When the UV radiation breaks oxygen-vacancy defect bonds in germanium-doped silica, electrons are set free and find their way to color center traps elsewhere in the glass structure. The new electron traps change the absorption properties of the doped silica in the UV portion of the spectrum. The positive net change in the absorption spectrum causes an increase in the refractive index.

The relief of induced stress and/or configuration changes in the glass structure of the fibre core when the bonds, photolytically broken by the radiation, may also play a significant role in changing the index of the glass. The amount of index change depends on several factors, such as the radiation conditions including wavelength, intensity, and total dosage of radiation; the glass composition in the fibre core; and the processing (such as hydrogen loading) of the fibre prior to radiation. In practice, the most commonly used light sources are KrF and ArF excimer laser might generate 248 and 193 nm optical pulses respectively. Strong photosensitivity with germanium-doped fibres has been observed at these two wavelengths where all the characteristics of FBG are pertinent in the development of FBG temperature sensor system.

3.6   Fabrication of Fibre Bragg Grating

The fabrication of FBG is formed by KrF excimer laser (248 nm wavelength), mask aligner, TLS, and OSA. TLS provides a light source which passes through the optical fibre while OSA plays a critical role in demodulation to detect the growth of fibre grating and obtain the relevant spectrum. The schematic diagram for the fabrication of FBG is shown in Fig. 3.7.

The jacket or cladding of the section where the grating is supposed to be written should be removed before the fibre is placed on the platforms. For a QPS fibre, which has a standard diameter of 125 μm, the fibre jacket should be removed by a cleaver or stripper. For other types of fibres, a mixture of 50% dichloromethane and 50% acetone could be used for the purpose. Alternatively, one could use Nitromors (a paint stripper) but it will take a longer time. For certain jacket materials, the percentage of dichloromethane and acetone in the mixture would need to be changed. When the fibre is placed on the platform, a slight strain could be applied to ensure that the fibre is straight. It is also noteworthy that the naked photosensitive fibre should be cleaned thoroughly with acetone or alcohol before placing it on the platforms, otherwise the UV beam can ablate any remaining jacket and might damage the phase mask.

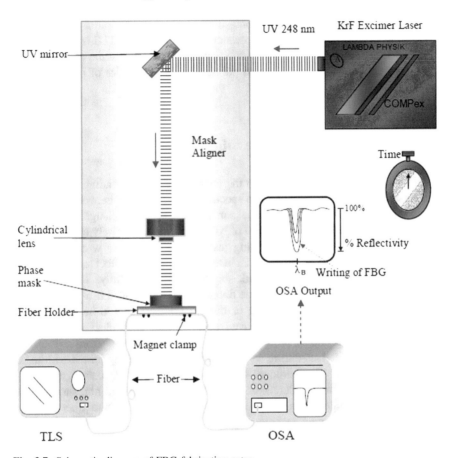

Fig. 3.7 Schematic diagram of FBG fabrication setup

The fibre is connected to the OSA and TLS. The real-time growth of the FBG is monitored with the OSA. It is necessary to clean all the optical elements in the mask aligner (reflecting mirror, cylindrical lens, phase mask, and quartz plate) with compressed nitrogen gas because any dust could absorb UV light and thus reduce the efficiency of the process. To ensure that the energy status of the excimer laser is suitable, several laser pulses with energy between 100 and 130 mJ at 20–30 kV voltage supply are tested in a closed tube. If the excimer laser output drops below the operation range, a new gas fill would be needed. After every new fill, a fine tune on the optical alignment of the laser pulses would be required.

With the completion of the optical alignment, the excimer laser is tuned to the pulsed mode. A suitable number of pulses are inserted to the laser controller and the grating writing process is started. The laser beam from the excimer laser enters the mask aligner and hits the fibre via the phase mask after passing through some mirrors and lens. The growth of the grating in terms of the center wavelength and reflectivity is monitored using OSA and the UV light that passes through the cylindrical lens is focused linearly onto the fibre core. To monitor the growth of the grating in transmission with a stopwatch, the light from the TLS is launched into the core of this fibre at one end and monitored with the OSA at the other end. The writing process is then stopped when the desired characteristics of the grating are achieved.

3.7   Preparation of No-Core Fibre

NCF exhibits a sensitive response to the changes of its surrounding environment, and the fabrication requires splicing a specific section of MMF between two SMFs. The length of NCF will influence the transmission peak wavelength. To gain the different peak wavelengths, the length of NCF needs to be precisely controlled. The fabrication of NCF will be done using fusion splicer. Splicing work is the most crucial stage of the fabrication of NCF since this process will affect the whole result of the experiment. The splicing work should be done carefully. The jacket and cladding of the single- and multi-mode fibres is strip or removed using optical fibre stripper. After this process, the bare or core of the fibre should be cleaned with alcohol. The fibre is then been cleaved using optical fibre cleaver. During the process, the cleaved end of the optical fibre should keep away from any dust to avoid causes of defectives during splicing work. It is then being joined with MMF using fusion splicer. The purposes of this stage is to ensure that light passing through the fibres is not scattered or reflected back. The cleaved fibres are aligned and then the adequate buffing pressure is applied, as per exact requirement. The losses due to the splicing technique should less than 0.05 dB and with proper aligned and care, this may be reduce to 0.00 dB loss.

Chapter 4
Methodology

4.1 Design of Fibre Bragg Grating for Temperature Sensing

A commercial FBG with a center wavelength of 1553.865 nm, 0.24 nm bandwidth, and >97% reflectivity was used for the purpose. It was used to investigate the thermal response by determining its sensitivity and accuracy. The indoor temperature (laboratory) system was increased by applying the heat of liquids. For the FBG sensor system in outdoor environment, sunlight was used as the natural temperature source. Figure 4.1 shows the general schematic diagram of the FBG temperature sensor system. The simplest system consisted of a light source (e.g., tunable laser source), a spectrum analyzer (e.g., optical spectrum analyzer), and an FBG. For indoor measurement, different types of liquid were used, which were distilled water and 0.1 M NaCl solution. Different focusing elements were used for the FBG sensor measurements in outdoor area, which were convex and hand lens with focal lengths of 20 and 15 cm, respectively.

4.2 Set Up of Fibre Bragg Grating for Indoor Temperature Sensor

Research and studies on the characteristics of FBG as a temperature sensor in liquid have been discussed earlier in Chap. 3. The experimental setup for temperature measurement in the liquid was formed up by a TLS, OSA, hot plate (for heating purposes), FBG, mercury-in-glass thermometer (to measure the temperature of liquid), and beaker filled with distilled water and 0.1 M NaCl solution. The schematic diagram of the experimental setup for temperature measurement in liquid is depicted in Fig. 4.2.

© The Author(s) 2018
S. Daud and J. Ali, *Fibre Bragg Grating and No-Core Fibre Sensors*,
SpringerBriefs in Physics, https://doi.org/10.1007/978-3-319-90463-4_4

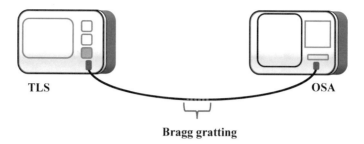

Bragg gratting

Fig. 4.1 Schematic diagram of basic FBG temperature sensor system

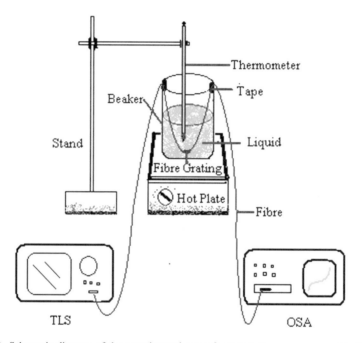

Fig. 4.2 Schematic diagram of the experimental setup for temperature measurement in liquid

FBG was connected to the TLS and OSA and placed in liquid solutions inside the beaker as shown in Fig. 4.3. A thermometer was placed in the beaker to record the temperature change of the solutions along the heating process. The temperature of solutions varied from 25 to 100 °C, thus producing the thermal expansion in the grating, which engendered the refractive index variation of the FBG. For the reasons, the laser source was turned on to operate this experiment. The Bragg wavelength of FBG was observed using OSA. All transmission and reflection spectra were recorded. Moreover, other critical properties such as bandwidth, power dip, and reflectivity of FBG were also recorded and investigated.

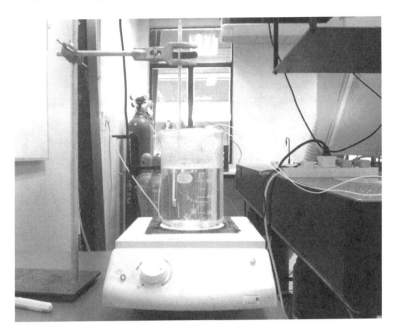

Fig. 4.3 Photograph of the experimental setup for temperature measurement in liquid

4.3 Set Up of Fibre Bragg Grating for Outdoor Temperature Sensor

Fibre Bragg grating has become very useful and important in sensing and telecommunication technologies. In this research, the FBG was placed in an open area and directly exposed to sunlight. Both transmission and reflection spectra were measured and analyzed. It is important to have both transmission and reflection measurements especially to cross-check the Bragg wavelength shift and the sensitivity of FBG for both transmission and reflection systems.

An FBG was inserted into a glass tube which is used to hold it at a certain height from the base of the perspex holder. This will prevent the FBG from being easily snapped and damaged (due to wind or rain). A thermometer was placed in the perspex holder and used to measure the surrounding temperature. The FBG was then directly connected with TLS and OSA through a single-mode fibre optic cable. A light source from TLS was launched into the core of the fibre and transmitted through FBG. As the light propagates through FBG, the light source with a wavelength that matches the Bragg condition will be reflected, while the rest will be transmitted through the fibre. These induce a significant power dip at the Bragg wavelength. The output of transmission spectrum was examined using OSA in terms of Bragg wavelength shift, bandwidth, and power dip. Figure 4.4 shows the schematic diagram of the experimental setup for measuring the transmission spectrum.

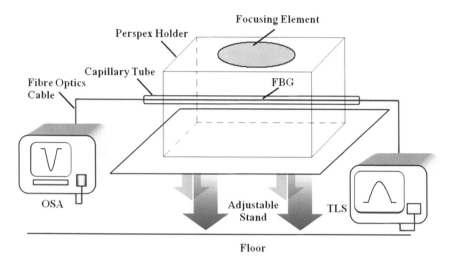

Fig. 4.4 Schematic diagram of experimental setup for transmission spectrum measurement

In order to achieve the reflection spectrum, a fibre optic coupler was added to the system. An SMF-28 2 × 2 3-dB fibre optic coupler was chosen. One end of the fibre optic coupler was connected to one end of the FBG, and another two ends were connected to the laser source (TLS) and output device (OSA). The schematic diagram of the experimental setup for examining the reflection spectrum is depicted in Fig. 4.5.

The reflection spectrum of FBG can be captured by using the OSA since light from TLS having the same range of wavelength with the FBG was launched into the fibre optic cable, to the FBG, and through the fibre optic coupler. One end of the FBG was neglected to ensure only the reflected spectrum will be measured by OSA. When the wavelength of the light from TLS matched with the Bragg wavelength of the FBG, it was reflected back and emerged as the output signal at OSA.

4.4 Set Up of Fibre Bragg Grating and No-Core Fibre for Temperature Sensing

In this part, the FBG and NCF were placed in three different types of solutions with various temperatures. The FBG and NCF used in this experiment were inserted in a 1 L solution (H_2O, NaCl, and NaOH). The FBG and NCF were dipped into the solution and directly connected with the TLS and OSA through a fibre optic cable. The sectional area of FBG and NCF was placed hanging in the beaker without touching the wall of the beaker to avoid overheating.

The light from TLS was launched into the core of the fibre and transmitted through the FBG and NCF. As the light propagates through both FBG and NCF, the

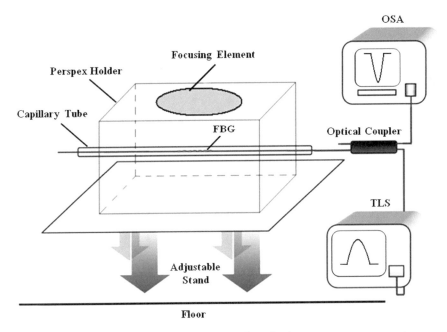

Fig. 4.5 Schematic diagram of experimental setup for reflection spectrum measurement

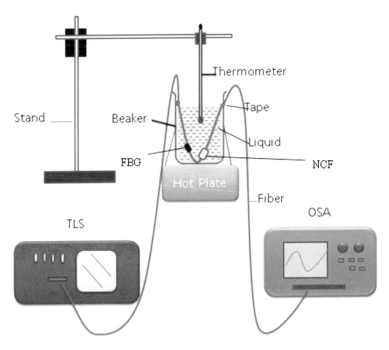

Fig. 4.6 Schematic diagram of the experimental setup for temperature measurement using FBG and NCF in liquid

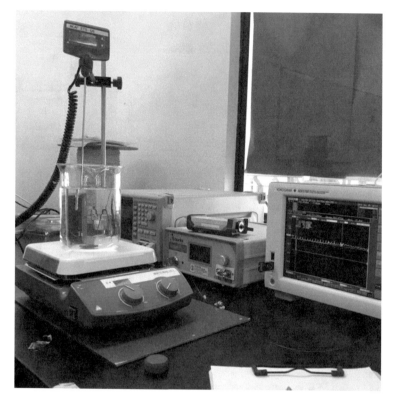

Fig. 4.7 Experimental setup for FBG and NCF temperature sensing

light source with a wavelength that matches the Bragg condition will be reflected, while the rest will be transmitted through the fibre. A significant power dip at the Bragg wavelength will be induced. OSA will display the output of the transmission spectra of both FBG and NCF in terms of Bragg wavelength, bandwidth, and power dip. Figure 4.6 shows the schematic diagram of the setup for measuring the transmission spectrum temperature. The experimental setup for FBG and NCF temperature sensing is shown in Fig. 4.7.

Chapter 5
Research Findings

5.1 Indoor Fibre Bragg Grating Temperature Sensor

5.1.1 Reflection and Transmission Spectra in Room Temperature

The manufactured spectrum of fibre Bragg grating (FBG) is presented in Fig. 5.1. It was used as the initial/reference reading of every measurement, for both transmission and reflection spectra. The center wavelength or the Bragg wavelength, λ_B, was measured as 1553.865 nm with 0.24 nm bandwidth. Figures 5.2 and 5.3 show the transmission and reflection spectra of FBG taken at room temperature (23 ± 1 °C).

The transmission spectrum (1553.44 nm wavelength, 0.38 nm bandwidth) and reflection spectrum (1553.42 nm wavelength, 0.37 nm bandwidth) are experimentally measured. The measured and the manufacturer's λ_B values show an excellent agreement with each other. Based on the data inspected, it is clear that the FBG can be used to measure both transmission and reflection spectra.

5.1.2 Fibre Bragg Grating for Temperature Sensor in Indoor Environment

The performance of FBG for the temperature sensor measurement in liquids was investigated in an indoor environment. For this reason, two different types of liquids were chosen, namely distilled water and 0.1 M NaCl solution. This was used to determine the effect of refractive index of the solution toward the performance of FBG. Tables 5.1 and 5.2 show data obtained from the experiments conducted in distilled water and NaCl solution respectively.

© The Author(s) 2018
S. Daud and J. Ali, *Fibre Bragg Grating and No-Core Fibre Sensors*,
SpringerBriefs in Physics, https://doi.org/10.1007/978-3-319-90463-4_5

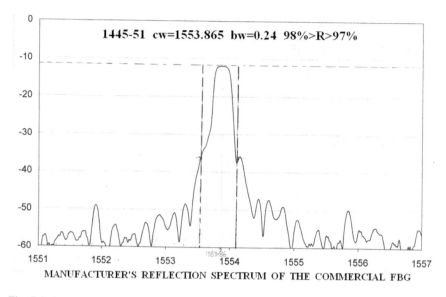

Fig. 5.1 Reflection spectrum provided by FBG manufacturer

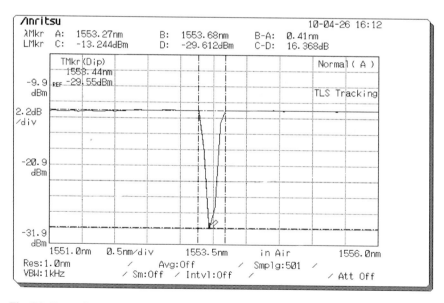

Fig. 5.2 Transmission spectrum at 23 °C

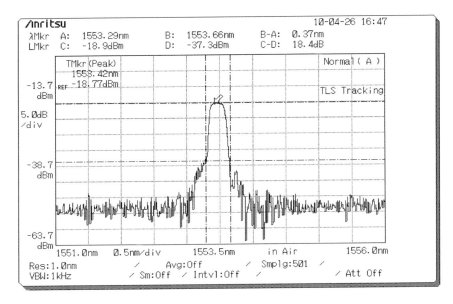

Fig. 5.3 Reflection spectrum at 23 °C

Table 5.1 Fibre Bragg grating for temperature measurement in distilled water

Temperature, T (± 1 °C)	Bragg wavelength, λ_B (± 0.01 nm)	Bandwidth, d (± 0.01 nm)	Power dip, r (± 0.005 dB)	Reflectivity, R (%)
25	1553.34	0.18	10.950	92
30	1553.38	0.18	1.8750	92
35	1553.46	0.18	10.950	92
40	1553.52	0.18	10.875	92
45	1553.58	0.18	10.950	92
50	1553.62	0.18	10.950	92
55	1553.65	0.18	10.950	92
60	1553.69	0.16	10.950	92
65	1553.73	0.18	10.950	92
70	1553.77	0.18	10.950	92
75	1553.81	0.18	10.950	92
80	1553.84	0.18	1.8750	92
85	1553.88	0.18	10.950	92
90	1553.92	0.18	10.875	92
95	1553.95	0.18	10.950	92
100	1553.97	0.18	10.950	92

Table 5.2 Fibre Bragg grating for temperature measurement in NaCl solution

Temperature, T (± 1 °C)	Bragg wavelength, λ_B (± 0.01 nm)	Bandwidth, d (± 0.01 nm)	Power dip, r (± 0.005 dB)	Reflectivity, R (%)
25	1553.28	0.18	10.220	92
30	1553.34	0.19	10.220	92
35	1553.38	0.18	10.220	92
40	1553.42	0.18	10.150	91
45	1553.50	0.18	10.220	92
50	1553.56	0.18	10.290	92
55	1553.60	0.18	10.290	92
60	1553.64	0.18	10.220	92
65	1553.68	0.18	10.220	92
70	1553.71	0.19	10.220	92
75	1553.75	0.18	10.220	92
80	1553.79	0.18	10.150	91
85	1553.84	0.18	10.220	92
90	1553.86	0.18	10.290	92
95	1553.91	0.18	10.290	92
100	1553.94	0.18	10.220	92

5.1.3 Bragg Wavelength

The shift in Bragg wavelength is dependent on temperature. The normalized wavelength change is related to temperature change by

$$\Delta\lambda_B = \lambda_B(\xi + \alpha)\Delta T \tag{5.1}$$

where $\Delta\lambda_B$ is the Bragg wavelength shift, λ_B is the Bragg wavelength of the FBG, ξ is the thermo-optic coefficient, α is the thermo-expansion coefficient, and ΔT is the change in temperature. For the germanium-doped silica core optical fibre, the values of ξ and α are 8.6×10^{-6} °C^{-1} and 0.55×10^{-6} °C^{-1} respectively.

From Eq. (5.1), with the constant values of ξ and α, it is mathematically shown that the shift in Bragg wavelength is dependent on temperature. An increase in temperature will increase the shift of Bragg wavelength and vice versa. It shows good agreement with the results obtained from the experiments done. With the increase in the temperature of the solutions, the Bragg wavelength of the fibre also increases. As a conclusion, the shift in Bragg wavelength is directly dependent on temperature.

Table 5.3 shows the comparison between theoretical results of Bragg wavelength shift and changes in temperature calculated using Eq. (5.1) and the data obtained using distilled water and NaCl solution. The graph of Bragg wavelength shift against temperature for theoretical calculation, distilled water, and NaCl solution is plotted and shown in Fig. 5.4. The graph shows that the Bragg wavelength behaved in an almost linear response to the changes in temperature.

Table 5.3 Theoretical, distilled water, and NaCl solution data for the shift in Bragg wavelength with temperature variation

Temperature, T (± 1 °C)	Change in T, ΔT (± 1 °C)	Bragg wavelength shift, $\Delta \lambda_B$ (± 0.01 nm)		
		Theoretical	Distilled water	NaCl solution
25	0	0	0	0
30	5	0.0711	0.08	0.10
35	10	0.1423	0.12	0.14
40	15	0.2134	0.16	0.16
45	20	0.2845	0.22	0.22
50	25	0.3557	0.26	0.28
55	30	0.4268	0.30	0.32
60	35	0.4979	0.38	0.36
65	40	0.5691	0.42	0.42
70	45	0.6402	0.44	0.48
75	50	0.7113	0.50	0.52
80	55	0.7825	0.54	0.56
85	60	0.8536	0.62	0.64
90	65	0.9247	0.68	0.70
95	70	0.9958	0.74	0.74
100	75	1.0670	0.78	0.78

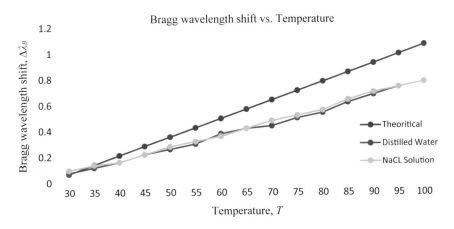

Fig. 5.4 Graph of Bragg wavelength shift versus temperature

This means that the Bragg wavelength shift, $\Delta \lambda_B$, is proportional to the change in the temperature. The data of theoretical and experimental results obtained were plotted for comparison. The data trend shows that the experimental results obtained matched with the theoretical calculation done.

The example of spectral results for Bragg wavelength measurement in this experiment is shown in Fig. 5.5. The center wavelength, reflectivity, power dip, and full width at half maximum ($\Delta\lambda_{FWHM}$) of the spectrum were measured.

The change in Bragg wavelength was calculated by deducting the initial Bragg wavelength from the second Bragg wavelength:

$$\Delta\lambda_B = \lambda_2 - \lambda_B \tag{5.2}$$

where $\Delta\lambda_B$ is the Bragg wavelength change, λ_2 is the second wavelength (or the Bragg wavelength at current temperature), and λ_B is the initial wavelength (or Bragg wavelength).

Figure 5.6 shows the examples of the easiest way to measure the Bragg wavelength shift with the change in temperature. The red line represents the spectrum at initial temperature, λ_B, and the green line represents the spectrum at current temperature, λ_2. As depicted in this figure, the values of λ_B and λ_2 were recorded as 1549.96 and 1550.28 nm respectively, and the shift in the Bragg wavelength is calculated as 0.32 nm or 320 pm. It was observed that there is a slight change of the Bragg wavelength with changes in temperature.

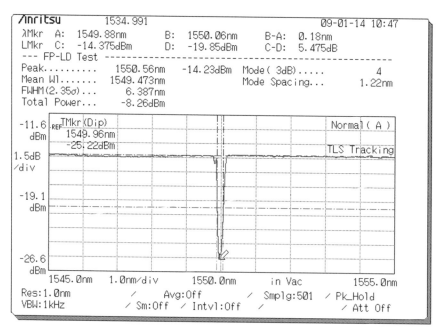

Fig. 5.5 Example of spectral results for Bragg wavelength shift

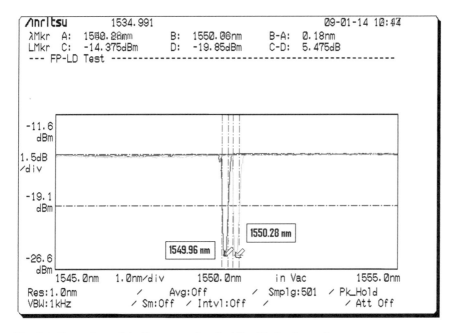

Fig. 5.6 Comparison of the Bragg wavelength shift with the change in temperature

5.1.4   Bandwidth

Measuring the full width at half maximum, $\Delta\lambda_{\text{FWHM}}$, is the simplest method used to determine the bandwidth of FBG. The bandwidth of FBG is dependent on two main factors, which are the number of grating plane, N and the depth of index modulation, Δn. Theoretically, the bandwidth of FBG can be determined using

$$\Delta\lambda_{\text{FWHM}} = \lambda_\text{B}s\sqrt{\left(\frac{\Delta n}{2n_\text{o}}\right)^2 + \left(\frac{1}{N}\right)^2} \tag{5.3}$$

Data taken from experiments shows that the values of the bandwidth of the FBG collected on FBG sensor in distilled water and NaCl solution are close to each other and it can be concluded that they are constant. There are only small changes and no dramatic changes in the bandwidth in neither distilled water nor NaCl solution measurements. The values of the bandwidth in distilled water are collected in the range of 0.16–0.18 (±0.01) nm. It is coherent with the results measured in NaCl solution where it was recorded as 0.18–0.19 (±0.01) nm.

The difference of bandwidth measured in these two experiments is very small, that is, 0.03 nm or 3×10^{-11} m. The bandwidth measured is almost similar when compared with the actual values discussed in Sect. 5.1.1 earlier. Table 5.4 shows the data collected for the bandwidth of FBG measured in distilled water and NaCl

Table 5.4 Bandwidth of FBG measured in distilled water and NaCl solution with temperature variation

Temperature, T (± 1 °C)	Bandwidth, d (± 0.01 nm)	
	Distilled water	NaCl solution
25	0.18	0.18
30	0.18	0.19
35	0.18	0.18
40	0.18	0.18
45	0.18	0.18
50	0.18	0.18
55	0.18	0.18
60	0.16	0.18
65	0.18	0.18
70	0.18	0.19
75	0.18	0.18
80	0.18	0.18
85	0.18	0.18
90	0.18	0.18
95	0.18	0.18
100	0.18	0.18

solution. This observation may lead to the conclusion that the values of bandwidth measured are unaffected by the changes in temperature and the refractive index of the solutions.

5.1.5 Reflectivity

The reflectivity of FBG can be calculated by substituting the value of power dip in Eq. (5.4)

$$R = \left(1 - 10^{-\frac{r}{10}}\right) \times 100 \tag{5.4}$$

where R is the reflectivity and r is the power dip of the spectrum.

Table 5.5 shows the data taken from experiments conducted for the reflectivity measurement of the FBG in distilled water and NaCl solution. It can be seen that the reflectivity of the FBG in distilled water and NaCl solution were almost constant even with changes in temperature. The reflectivity of the FBG in distilled water is measured to be 92% while in NaCl solution, it is measured to be in the range from 90 to 91%. These values coincide with the actual reflectivity value of FBG, which is 91%. It can be concluded that there is no variation in reflectivity with the variation in temperature, either in distilled water or NaCl solution. The reflectivity of FBG is independent of the change in temperature and refractive index of the solutions.

Table 5.5 Reflectivity of FBG in distilled water and NaCl solution

Temperature, T (± 1 °C)	Change in T, ΔT (± 1 °C)	Reflectivity, R (%)	
		Distilled water	NaCl solution
25	0	92	91
30	5	92	91
35	10	92	91
40	15	92	91
45	20	92	91
50	25	92	90
55	30	92	91
60	35	92	90
65	40	92	91
70	45	92	91
75	50	92	91
80	55	92	90
85	60	92	91
90	65	92	91
95	70	92	91
100	75	92	91

5.1.6   Fibre Bragg Grating Sensor in Liquid

FBG characteristics (Bragg wavelength, bandwidth, and reflectivity) can be linked with temperature variation. The Bragg wavelength is proportional to the temperature change while the bandwidth and reflectivity of the FBG are always constant with changes in temperature and refractive index of the solutions. In other words, bandwidth and reflectivity of FBG are independent of the temperature and refractive index of the solutions.

5.1.6.1   Sensitivity of Fibre Bragg Grating in Liquid

The sensitivity of fibre is defined as the ability to respond for affective changes in certain environments, which implies the susceptibility of a system to the changes surrounding it. Thus, temperature sensitivity of the FBG sensor was investigated in terms of how sensitive the sensor is toward temperature change. In principle, lower sensitivity indicates better susceptibility.

Figures 5.7 and 5.8 show the graph of Bragg wavelength shift versus temperature for the FBG sensing system in distilled water and NaCl solution. The slope steepness of the graph determines the sensitivity and minimum detectable wavelength shift of the sensor. The average sensitivity of FBG can be obtained by calculating the countable slope steepness of the graph plotted. From the graphs

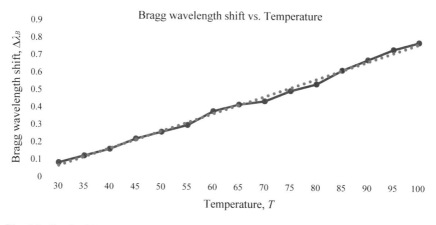

Fig. 5.7 Graph of Bragg wavelength shift versus temperature in distilled water

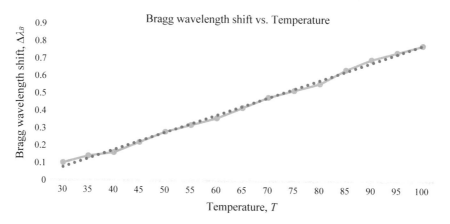

Fig. 5.8 Graph of Bragg wavelength shift versus temperature in NaCl solution

plotted, the Bragg wavelength shift is directly proportional to the change in temperature for both systems. The sensitivities of FBG in distilled water and NaCl solution were measured to be 0.0109 and 0.0102 nm $°C^{-1}$ respectively. Nevertheless, the FBG sensor has been proven as a usable candidate for temperature sensing.

5.1.6.2 Resolution and Accuracy of Fibre Bragg Grating Temperature Sensor

The resolution of the temperature sensor is the smallest change it can detect in the temperature measured. By knowing the wavelength resolution from the sensitivity obtained, the temperature resolution can be determined as well. Standard deviation

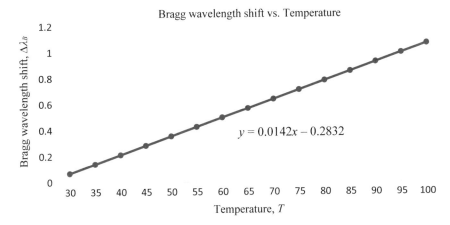

Fig. 5.9 Theoretical sensitivity of FBG temperature sensor

is used to indicate the accuracy of any devices. The accuracy of the fabricated FBG temperature sensor can be affirmed by examining its standard deviation. Figure 5.9 shows the graph of Bragg wavelength shift versus temperature for the theoretical calculation of the FBG temperature sensor. From the graph plotted, the sensitivity of FBG was measured to be 0.0142 nm $°C^{-1}$ and intercepted at 0.2832 nm wavelength.

Based on the testing done in different solutions, it is measured that the temperature sensitivity of FBG sensor in distilled water and NaCl is 0.0109 and 0.0102 nm $°C^{-1}$ respectively, whereas the theoretical temperature sensitivity is measured to be 0.0142 nm $°C^{-1}$. Based on this situation, it can be concluded that the FBG may work as a viable temperature sensor with a sensitivity of 0.0142 nm $°C^{-1}$ and resolution of 0.0065 $°C$.

5.1.7 Fibre Bragg Grating Temperature Sensor in Liquid and Air

FBG act as sensors not only in liquids, but also in air, gases, as well as corrosion studies. Table 5.6 shows the data of FBG characteristics in distilled water and air collected for the comparison. The same FBG is used for the purpose. The Bragg wavelength shift of FBG in liquids increases with an increase in temperature as well as in the air. Also, the bandwidth and reflectivity of FBG in the air do not change with the changes in temperature and refractive index of the medium.

Table 5.6 Comparison of the characteristics of FBG in distilled water and air

Temperature, T (±1 °C)	Distilled water			Air		
	Bragg wavelength, λ_B (±0.01 nm)	Bandwidth, d (±0.005 nm)	Reflectivity, R (%)	Bragg wavelength, λ_B (±0.01 nm)	Bandwidth, d (±0.005 nm)	Reflectivity, R (%)
25	1549.84	0.180	92	1550.06	0.180	92
35	1549.96	0.180	92	1550.14	0.180	92
45	1550.06	0.180	92	1550.18	0.180	91
55	1550.14	0.180	92	1550.24	0.180	91
65	1550.26	0.180	92	1550.30	0.180	91
75	1550.34	0.180	92	1550.38	0.190	92
85	1550.46	0.180	92	1550.44	0.180	91
95	1550.58	0.180	92	1550.56	0.180	91
100	1550.64	0.180	92	1550.63	0.180	91

5.2 Outdoor Fibre Bragg Grating Temperature Sensor

5.2.1 Fibre Loss Measurement

In the research proposed for outdoor temperature environments, 1550 nm single-mode fibre (SMF) system of 55.5 m length (consisting of two pieces of 26.0 m SMF-28 fibre and 3.5 m FBG sensor) were used. Both ends of the fibre were connected to the FC-FC connector (so that it matches with FBG, TLS, OSA, and fibre optic coupler). The measurements were set up from TLS via FBG to OSA and from TLS to FBG via fibre optic coupler, to OSA.

In performing the experimental measurement of fibre loss, an initial reading was taken when one meter optical fibre was connected with a light source and power meter. It was then connected with a 26-m optical fibre which was used in this research, and a second reading was measured. Data taken from both readings are recorded in Table 5.7. The difference between the final and initial reading is the loss of that 26-m fibre. In this research, the loss of 26-m optical fibre was 4.21 dB, leading to an 8.42-dB loss in the full FBG outdoor temperature sensor system.

Determination of optical losses in the system used is important indeed. Assuming that the total optical loss in the system is the summation of connector loss and attenuation loss leads to

$$\sum \text{Losses} = \text{Connector loss} + \text{Attenuation loss} \tag{5.5}$$

The loss of the connector is 2.0 dB each. Thus, the total connector loss is

$$4(\text{connectors}) \times (2.0)\,\text{dB} = 8.0\,\text{dB} \tag{5.6}$$

And the attenuation coefficient of the system can be measured using

$$P_{\text{out}} = P_{\text{in}}e^{-\alpha L} \tag{5.7}$$

α in Eq. (5.7) is the attenuation coefficient and it is given as

$$\alpha = \frac{10}{L}\log_{10}\left(\frac{P_{\text{in}}}{P_{\text{out}}}\right) \tag{5.8}$$

where L is the fibre length and P_{in} and P_{out} are the input and output powers of the system respectively. The values of P_{in} and P_{out} were measured using OSA in the course of this experiment. P_{in} and P_{out} are given as

Table 5.7 Power meter reading of fibre loss	Fibre length, l (±0.1 m)	Output power, P_{out} (±0.01 dB)
	1.0	37.16
	27.0 (1.0 + 26.0)	32.95
	Loss	4.21

$$P_{in} = -10.00\,dBm = 4.71 \times 10^{-7}\,W = -63.27\,dB \tag{5.9}$$

$$P_{out} = -33.27\,dBm = 10.00 \times 10^{-4}\,W = -30.00\,dB \tag{5.10}$$

The total length of the system, L, can be calculated using

$$L = 26.0\,m + 26.0\,m + 3.5\,m = 55.5\,m \tag{5.11}$$

Using Eq. (5.8), the attenuation coefficient is equal to -0.0584 dB km^{-1}.
    Thus, the attenuation loss is

$$4.34\,\alpha = 0.2534\,dB \tag{5.12}$$

By using Eq. (5.5), the total loss of the FBG temperature sensor system is calculated to be 8.2534 dB.
    The total loss investigation and calculation obtained from experimental results and calculation measurements show that the results were in good agreement with each other.

5.2.2   Fibre Bragg Grating for Temperature Sensor in Outdoor Environment

Research on FBG for temperature sensing in an open area was conducted. The performance of FBG for different placement heights was measured during different periods of the day with and without any focusing element. The FBG sensor head was focused with a convex and hand lens to measure the effect of focusing elements toward the performance of FBG sensor. The effect of the focusing elements was identically measured and analyzed. Figure 5.10 shows typical examples of transmission and reflection spectra for different temperatures in an outdoor environment.
    The Bragg wavelength shift is the difference between the Bragg wavelength measured at a certain temperature and the Bragg wavelength measured at room temperature (also called initial temperature).
    It is calculated using:

$$\Delta\lambda_B = \lambda_{B,T} - \lambda_B \tag{5.13}$$

where $\Delta\lambda_B$ is the Bragg wavelength shift, $\lambda_{B,T}$ is the Bragg wavelength measured at temperature T, and λ_B is the Bragg wavelength measured at room temperature of 23 °C.
    An analysis of Bragg wavelength shift of transmission and reflection spectra were performed over outdoor temperature variations ranging from 23 to 42 °C respectively. Figure 5.10 shows the typical FBG transmission and reflection spectra due to the effect of outdoor temperatures obtained via OSA. The dotted line is used as an indicator to show the shift in Bragg wavelength. In these experiments, the Bragg wavelength shift illustrated in Fig. 5.11 is based on the transmission spectrum.

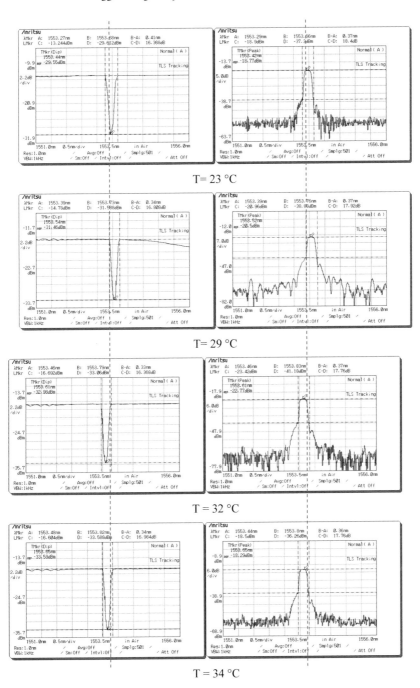

Fig. 5.10 Examples of transmission and reflection spectra

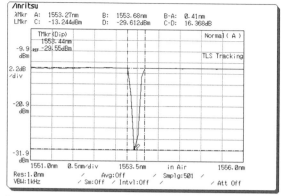

$\lambda_B = 1553.44$ nm $T_o = 23$ °C

$\Delta\lambda_{B1} = \lambda_{B1} - \lambda_B$
where λ_B is the
Bragg wavelength at
room temperature,
T_o, and λ_{B1} is the
Bragg wavelength at
T_1.

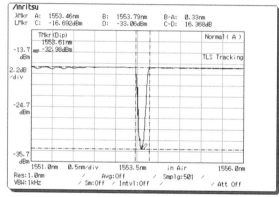

$\lambda_{B1} = 1553.61$ nm $T_1 = 32$ °C

$\Delta\lambda_{B2} = \lambda_{B2} - \lambda_B$
where λ_B is the
Bragg wavelength at
room temperature,
T_o, and λ_{B2} is the
Bragg wavelength at
T_2.

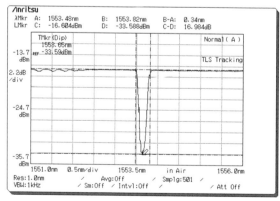

$\Delta\lambda_{B2} = 1553.65$ nm $T_2 = 34$ °C

Fig. 5.11 Calculation of Bragg wavelength shift

Typical wavelength shifting response of FBG sensor due to temperature can be achieved by calculating the gradient, assuming constant strain. Thus, temperature sensitivity with good linear characteristics over practical dynamic ranges can be obtained. Experiments were performed over a period of time to achieve different effects, namely placement heights, temperature variations, and effect of focusing lens based on the transmission and reflection spectra characteristics. These spectra provided valuable information on the Bragg wavelength shift, power dip, bandwidth, reflectivity, and transmissivity of the FBG at different temperatures.

5.2.3 Height Effect on the Sensitivity of Fibre Bragg Grating

Does the elevated position of FBG affect its sensitivity? Figure 5.12 shows the graph of Bragg wavelength shift versus temperature for the FBG sensor head when it is placed at different heights based on the transmission spectrum. From the graph plotted, the profile shows a linear response, where the gradient of the profile determines the sensitivity of FBG. Similarly, Fig. 5.13 shows the effect of outdoor temperature variations on Bragg wavelength shift. The graph of Bragg wavelength shift versus temperature plotted shows a linear relationship between the two variables. Table 5.8 shows the summary of the sensitivity of FBG calculated based on both transmission and reflection spectra for different placement heights.

The average sensitivities measured for the transmission and reflection spectra are 10.05 and 10.02 pm $°C^{-1}$ respectively. Based on the data taken and graph plotted, the results show an excellent agreement for the sensitivity of FBG for both systems measured. This shows that the sensitivity of FBG is not affected by its placement height. The results also show that the FBG sensitivity can be determined either from transmission or reflection spectra.

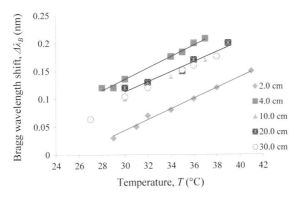

Fig. 5.12 Bragg wavelength shift in different height for transmission spectrum

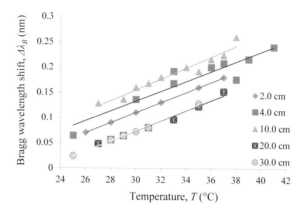

Fig. 5.13 Bragg wavelength shift at different heights for reflection spectrum

Table 5.8 Sensitivity of FBG at different heights

Height, h (± 0.1 cm)	Sensitivity (pm °C^{-1})	
	Transmission spectrum	Reflection spectrum
2.0	10.00	10.00
4.0	10.09	10.00
10.0	10.15	10.00
20.0	10.19	10.00
30.0	9.80	10.10
Mean	10.05	10.02

5.2.4 Focusing Lens Effect on the Sensitivity of Fibre Bragg Grating

Different types of focusing element will result in a different output of the system. A research was done to measure if either of the focusing elements affect the sensitivity of FBG. A convex lens of 20 cm focal length and a hand lens of 15 cm focal length were used in this research for the purpose. All the focusing elements were placed directly on top of the sensor head of FBG.

5.2.4.1 Effect of Convex Lens

The setup depicted in Figs. 4.4 and 4.5 in Chap. 4 were prepared. Convex lens was chosen as the focusing elements for this purpose. Figures 5.14 and 5.15 show the graphs of Bragg wavelength shift versus temperature plotted for the transmission and reflection spectra with the presence of convex lens on top of the FBG's sensor

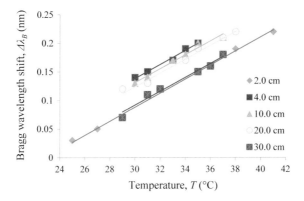

Fig. 5.14 Effect of convex lens on the Bragg wavelength in transmission system

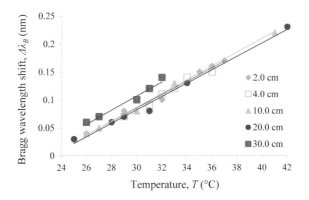

Fig. 5.15 Effect of convex lens on the Bragg wavelength in reflection system

Table 5.9 Effect of convex lens on the sensitivity of FBG at different heights

Height, h (± 0.1 cm)	Sensitivity (pm $°C^{-1}$)	
	Transmission spectrum	Reflection spectrum
2.0	12.22	12.27
4.0	12.00	12.40
10.0	12.73	12.39
20.0	12.14	12.35
30.0	12.32	12.37
Mean	12.28	12.35

head. Based on the data taken and a graph plotted, the Bragg wavelength shift is directly proportional to the change of temperature for both systems. The computed FBG sensitivity is summarized in Table 5.9.

The sensitivities of the FBG are calculated from the gradient of the graphs of Bragg wavelength shift against temperature as plotted in Figs. 5.14 and 5.15. The sensitivities of each graph plotted are calculated and the average sensitivity of the transmission and reflection system are recorded as 12.28 and 12.35 pm °C^{-1} respectively. Both transmission and reflection systems show an excellent agreement on the sensitivity of the FBG.

5.2.4.2 Effect of Hand Lens

The setup as discussed in Sect. 5.2.4.1 was repeated where a hand lens with 15.0 cm focal length was used to replace the convex lens. The measurements of Bragg wavelength shift based on transmission and reflection spectra were made for different placement heights of the FBG at different temperatures. Figures 5.16 and 5.17 show the graph of the Bragg wavelength shift versus temperature for both

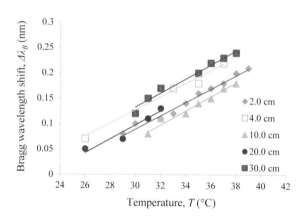

Fig. 5.16 Effect of hand lens on the Bragg wavelength in transmission system

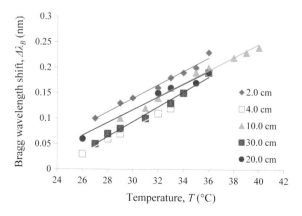

Fig. 5.17 Effect of hand lens on the Bragg wavelength in reflection system

Table 5.10 Effect of hand lens on the sensitivity of FBG at different heights

Height, h (± 0.1 cm)	Sensitivity (pm °C^{-1})	
	Transmission spectrum	Reflection spectrum
2.0	13.33	14.00
4.0	13.35	13.73
10.0	13.33	14.00
20.0	13.00	13.85
30.0	13.56	14.00
Mean	13.31	13.92

transmission and reflection systems where the FBG sensor head is focused directly with the hand lens.

From the graphs plotted in Figs. 5.16 and 5.17, it can be observed that the shift in the Bragg wavelength is directly proportional to the change in temperature. The sensitivities of each graph plotted e measured where an average sensitivity of FBG system is recorded as 13.31 and 13.92 pm °C^{-1} for the transmission and reflection systems respectively. The calculated sensitivities of FBG for this research are summarized in Table 5.10.

5.2.4.3   Comparison of Different Focusing Elements

As discussed in Sects. 5.2.4.1 and 5.2.4.2, the change in the Bragg wavelengths is directly proportional to the change in temperature. This applies to both the transmission and reflection systems. The presence of the focusing elements (convex and hand lens) affects the sensitivity of FBG sensor. Figures 5.18 and 5.19 show the

Fig. 5.18 Comparison of Bragg wavelength shift at 2.0 cm height

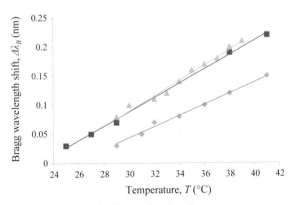

◆ FBG sensor head without a focusing element

■ FBG sensor head with a convex lens as the focusing element

▲ FBG sensor head with a hand lens as the focusing element

Fig. 5.19 Comparison of
Bragg wavelength shift at
30.0 cm height

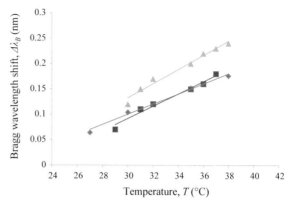

◆ FBG sensor head without a focusing element

■ FBG sensor head with a convex lens as the focusing element

▲ FBG sensor head with a hand lens as the focusing element

Fig. 5.20 Mean sensitivity of
FBG sensor in the unfocused
system

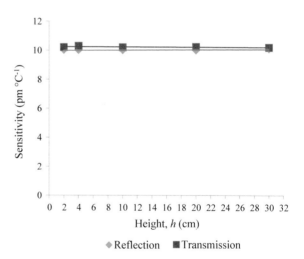

◆ Reflection   ■ Transmission

graph of Bragg wavelength shift against temperature at 2.0 and 30.0 cm heights
respectively with different focusing elements applied.

Figures 5.20, 5.21, and 5.22 show the average sensitivity of FBG in focused and
unfocused system modes based on the transmission and reflection spectra. The
placement height of FBG does not affect the sensitivity of the FBG sensor. The
sensitivities of the FBG are constant on both the transmission and reflection spectra
in unfocused modes and slightly increased in the focused modes but the values are
almost the same for both transmission and reflection spectra. The sensitivities of the
systems increase further when the convex lens is replaced with the hand lens. It can

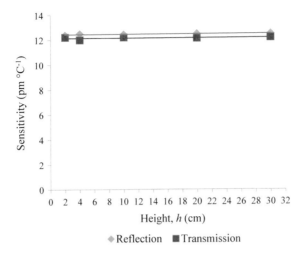

Fig. 5.21 Mean sensitivity of the FBG sensor focused using the convex lens

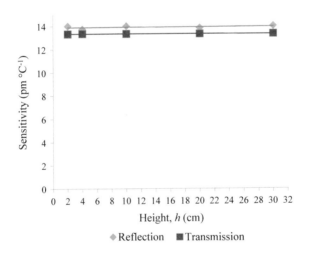

Fig. 5.22 Mean sensitivity of the FBG sensor focused using the hand lens

be concluded that the presence of a hand lens on the sensor head will increase the sensitivity of FBG sensor due to the thermal expansion of the grating length.

Tables 5.11 and 5.12 show the sensitivities of the FBG for the transmission and reflection systems respectively. Data taken from the experiment done without a focusing element, focused using a convex lens and focused using a hand lens are shown. The data discussed are based on actual measurements done in outdoor environments, under direct sunlight as the temperature source.

From the data tabulated in Tables 5.11 and 5.12, it is clear that the FBG sensor head focused using a hand lens has greater sensitivity compared to other systems.

Table 5.11 Sensitivity of the FBG for different focusing elements based on the transmission spectra

Height, h (±0.1) cm	Sensitivity (pm °C^{-1})		
	Without focusing element	Focus with convex lens	Focus withhand lens
2.0	10.00	12.22	13.33
4.0	10.00	12.00	13.35
10.0	10.00	12.73	13.33
20.0	10.00	12.14	13.00
30.0	10.10	12.32	13.56

Table 5.12 Sensitivity of FBG for different focusing elements based on the reflection spectra

Height, h (±0.1) cm	Sensitivity (pm °C^{-1})		
	Without focusing element	Focus with convex lens	Focus with hand lens
2.0	10.00	12.27	14.00
4.0	10.09	12.40	13.73
10.0	10.15	12.39	14.00
20.0	10.19	12.35	13.85
30.0	9.80	12.37	14.00

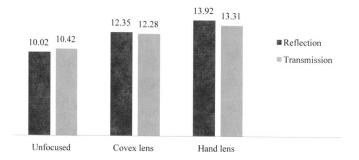

Fig. 5.23 Bar chart of the mean sensitivity of FBG sensor

The hand lens has a significant effect on the FBG's sensitivity for both transmission and reflection spectra. The comparison of both transmission and reflection spectra for all systems are presented in Fig. 5.23.

5.2.5 Bandwidth of Fibre Bragg Grating

Bandwidth is the separation of wavelength between two points on either side of the Bragg wavelength where the reflectivity has decreased to half of its maximum

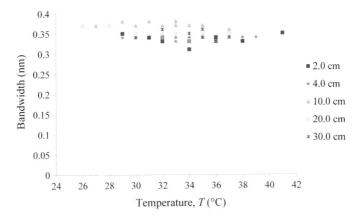

Fig. 5.24 Bandwidth of the FBG at different placement heights

value. Figure 5.24 shows the graph of bandwidth against the temperature profile for different FBG placement heights at different temperatures.

Referring to the data plotted in Fig. 5.24, the results show that the bandwidth of FBG is almost constant, even at different heights, with different focusing elements applied, and variations of temperature. The bandwidth of FBG is recorded in the range from 0.32 to 0.37 nm. It can be concluded that the variations in temperature and placement height do not affect the bandwidth of FBG. The value of FBG's bandwidth is always constant irrespective of any changes in temperature, focusing elements applied, and placement heights.

5.2.6 Reflectivity of Fibre Bragg Grating

When light is launched into the FBG, one center wavelength will be reflected back and the rest of wavelength will be transmitted. The reflectivity of the FBG can be calculated using Eq. (3.21), which is

$$R = \left(1 - 10^{-\frac{d}{10}}\right)100$$

where R is the reflectivity and d is the power dip of the spectrum.

Figure 5.25 shows the data collected for the reflectivity of FBG at different placement heights and temperature. It shows that the reflectivity of FBG is always constant for all placement heights of the FBG. The temperature variation does not affect the reflectivity as well. Different type of focusing elements applied does not affect the reflectivity of FBG. The reflectivity of FBG measured in this research is in the range of 86–88%.

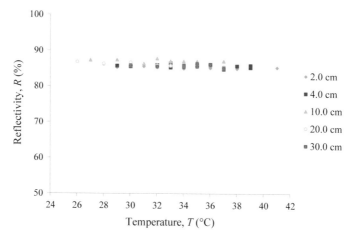

Fig. 5.25 Reflectivity of FBG at different height and temperature

5.2.7 *Interpretation and Discussion on Fibre Bragg Grating Sensor in Outdoor Environment*

The shifts in Bragg wavelength with external temperature applied are observed for both transmission and reflection spectra. This is due to the perturbations of gratings resulting in a shift in Bragg wavelength in either transmission or reflection systems. As the outdoor temperature changes due to environmental conditions, thermal expansion in the grating occurs. Due to this thermal expansion, the refractive index, n_{eff} of the FBG changes. This causes variation in the wavelength of FBG itself. The variations in $\Delta\lambda_B$ are monitored through the transmission or reflection spectra via the spectrum analyzer.

The reflection method offers some advantages over the transmission method. In the reflection system, only light that matches the Bragg condition of the grating is measured over a relatively small background intensity. There is a good correlation between Bragg wavelength shift and temperature changes obtained from the setups designed. A linear response is observed between Bragg wavelength shift and temperature changes throughout the measured region. The slope or gradient of $\Delta\lambda_B$ versus T describes the sensitivity of FBG sensors.

Results obtained from the graph of Bragg wavelength shift plotted against outdoor temperature show the sensitivities of 12.35 and 13.92 pm °C^{-1} for the sensor head focused using convex and hand lens respectively. In the unfocused mode system, the sensitivity of FBG was calculated to be 10.02 pm °C^{-1}. These values of FBG sensitivities differed for the focused and unfocused FBG sensing systems. FBG sensor head focused using the hand lens showed a higher sensitivity compared to the convex lens or without focusing element.

For an FBG with a 1550-nm center wavelength, the typical sensitivity is approximately 13 pm °C^{-1}. Theoretically, the change in Bragg wavelength, $\Delta\lambda_B$ for a given wavelength of 1553.805 nm due to the change in temperature, ΔT is calculated to be 13.2 pm $°C^{-1}$. It can be seen that there was a good agreement between the experimental results, the theoretical calculation, and the typical temperature sensitivity. It is concluded that the prototype FBG temperature sensing system can be used not only indoors but also outdoors.

5.3 Fibre Bragg Grating and No-Core Fibre for Temperature Measurement

5.3.1 Fibre Bragg Grating and No-Core Fibre at Room Temperature

The room temperature was measured and recorded. The reading was then used as the initial reading for the whole study of FBG and NCF temperature sensor. Figure 5.26 shows the FBG and NCF spectra taken at room temperature of 29 °C. The center wavelength of FBG spectrum is recorded as 1549.5600 nm with a

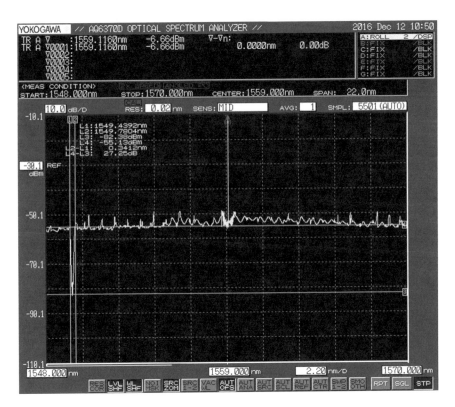

Fig. 5.26 FBG and NCF spectra at 29 °C

bandwidth of 0.34 nm. The power dip and FWHM of this spectrum are recorded as 27.25 dB and 0.1706 nm respectively and the center wavelength of no-core fibre, λ_{NCF}, is recorded as 1559.1160 nm.

5.3.2   Temperature Sensing of Fibre Bragg Grating and No-Core Fibre

The studies of FBG and NCF for temperature sensing were conducted in three different types of solutions, namely distilled water, NaCl, and NaOH solutions. The hot plate was used to supply heat to the systems, varied up to 100 °C. The performances of FBG and NCF at different temperatures were measured while the solutions were changed and the temperature was increased. Figures 5.27, 5.28, and 5.29 show the typical transmission spectra for the FBG and NCF in different solutions at a particular temperature.

The spectrum depicted in Fig. 5.27 shows that the center wavelength or Bragg wavelength, λ_B, is 1550.0760 nm with a bandwidth of 0.32 nm, while the power

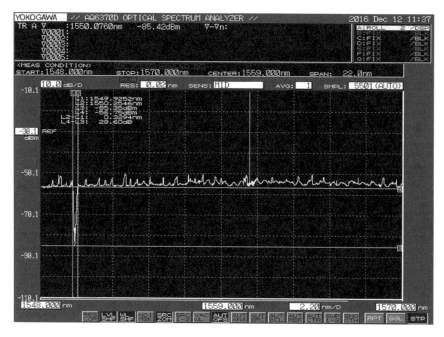

Fig. 5.27 FBG and NCF spectrum in H_2O at 90 °C

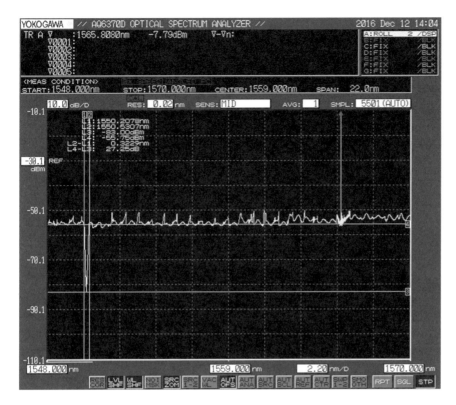

Fig. 5.28 FBG and NCF spectrum in NaCl solution at 90 °C

dip and FWHM are recorded as 28.6 dB and 0.1647 nm respectively. The center wavelength of NCF, λ_{NCF}, is 1560.724 nm. Details of λ_B, λ_{NCF}, bandwidth, power dip, and FWHM of the systems are presented in Table 5.13.

An analysis of the Bragg wavelength shifts of the fibre is performed for different solutions and temperatures from 30 to 100 °C. Figure 5.30 shows the typical FBG and NCF spectra at various temperatures collected from the spectrum analyzer. The figure illustrates the Bragg wavelength and no-core fibre shift in NaCl solution at temperatures 30, 60, and 100 °C, respectively.

Figure 5.30 shows the small shifts in the wavelength of FBG spectrum at different temperatures of the NaCl solution. The wavelengths recorded are 1549.630, 1549.880, and 1550.356 nm at 30, 60, and 90 °C, respectively. A bigger shift in the wavelength of NCF is recorded, and the spectra show 1556.590, 1562.350, and 1565.808 nm at 30, 60, and 90 °C, respectively.

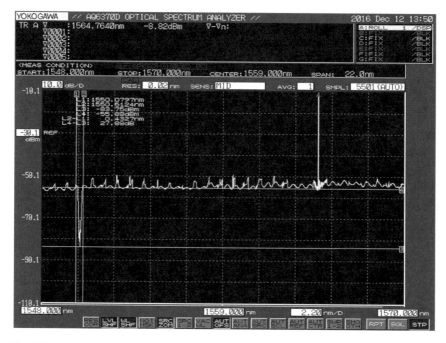

Fig. 5.29   FBG and NCF spectrum in NaOH solution at 90 °C

Table 5.13   Details of FBG and NCF spectra

Properties	Solution		
	H_2O	NaCl	NaOH
Bragg wavelength (nm)	1550.0760	1550.3560	1550.1880
NCF wavelength (nm)	1560.7240	1565.8080	1566.4560
Bandwidth (nm)	*0.32*	0.32	0.32
Power dip (dB)	28.60	27.25	30.48
FWHM (nm)	0.1647	0.1615	0.1615

5.3.3   Reaction of FBG in Different Solutions

The FBG and NCF were dipped into the solution at an initial temperature of 29 °C. The measurement for the spectrum was taken when the solution was heated. It was then repeated for different solutions starting from H_2O, NaCl, and NaOH solutions for temperatures ranging between 30 and 90 °C with an increment of 10 °C.

Figure 5.31 shows the graph of Bragg wavelength shift versus the temperature when the FBG sensor was placed in different solutions over increasing temperature over time. The profile shows a linear response whereas the gradient of the graph indicates the sensitivity of the FBG. From the graph plotted, an overall increasing

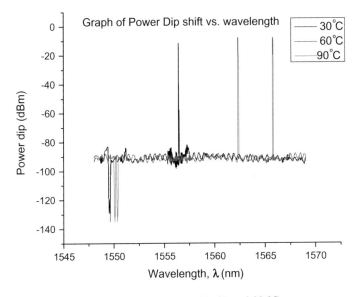

Fig. 5.30 Shifts of wavelength in NaCl solution at 30, 60, and 90 °C

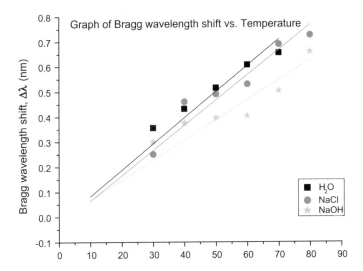

Fig. 5.31 Shift of Bragg wavelength in different types of solutions at different temperatures

trend can clearly be seen when the temperature increases for all types of solutions. Figure 5.32 shows the graph of the Bragg wavelength shift versus temperature of the theoretical and experimental values of the systems. The trend of the data shows that the experimental results match the theoretical calculation.

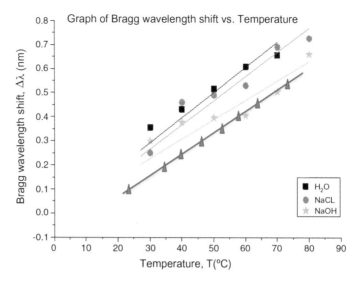

Fig. 5.32 Shift of Bragg wavelength in different types of solutions at different temperatures compared with the theoretical value of the system

5.3.4   Reaction of NCF in Different Solutions

Figure 5.33 shows the graph of no-core wavelength shift versus temperature when the NCF sensor was placed in different solutions with increasing temperature over time. The profile shows a linear response whereas the gradient of the graph

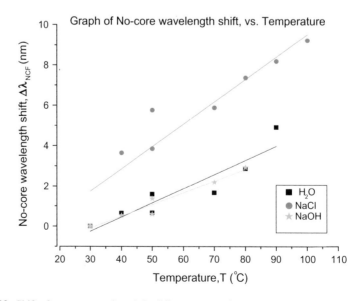

Fig. 5.33 Shift of no-core wavelength in different types of solutions at different temperatures

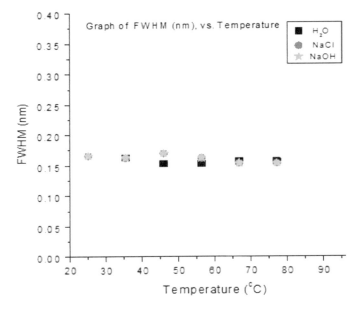

Fig. 5.34 FWHM versus temperature in different solutions

indicates the sensitivity of the NCF. From the graph plotted, a trend of significant increase is observed for the no-core wavelength shift as the temperature increases for all the solutions used. As for H_2O, NaCl, and NaOH solutions, the deviations of points along the line of best fit are not more than ± 1.21 nm.

5.3.5   Full Width at Half Maximum of Bragg Grating

The full width at half maximum (FWHM) is the difference between the two extreme values of the wavelength in which the dependent variable is equal to half of its maximum value. In other words, it is the width of a spectrum curve measured between those points on the y-axis; is half of the maximum amplitude. The FWHM of FBG is determined from the spectra shown in the spectrum analyzer. Figure 5.34 shows the graph of FWHM versus temperature profile in different solutions at different temperatures. From the graph plotted, it shows that the temperature of each solution does not affect the reading of the FWHM. The trend for FWHM of FBG is constant as the distribution of points of FWHM is consistent, ranging from 0.15 to 0.20 nm at a temperature ranging from 30 to 90 °C.

Chapter 6
Conclusion

A prototype of the indoor and outdoor fibre Bragg grating temperature sensor incorporated with no-core fibre sensor systems were designed, developed, investigated, and evaluated successfully. A commercial FBG sensor head with 1553.865 nm center wavelength, 0.24 nm bandwidth, and 3.0 ± 0.1 cm length and 3.5 ± 0.1 cm NCF sensor were used for the purpose. The prototype FBG and NCF temperature sensor systems were capable of detecting the Bragg wavelength shifts and a no-core shift in different indoor and outdoor temperatures, in different liquids, and at different heights, either with or without the presence of focusing elements. Experimental results found a linear relationship between the Bragg wavelength shift, $\Delta\lambda_B$, and temperature, T, for both transmission and reflection systems either in indoor or outdoor measurements. This is also applied in NCF systems. Thus, a prototype FBG and NCF sensing systems for temperature sensors have been developed.

© The Author(s) 2018
S. Daud and J. Ali, *Fibre Bragg Grating and No-Core Fibre Sensors*,
SpringerBriefs in Physics, https://doi.org/10.1007/978-3-319-90463-4_6

References

Bhardwaj, V., & Singh, V. K. (2017). Study of liquid sealed no-core fiber interferometer for sensing applications. *Sensors and Actuators A., 254,* 95–100.

Bowei, Z. (2004). *High temperature sensors based on hydrogen loaded fiber Bragg gratings.* M. Sc. Thesis. Concordia University, Canada.

Bowei, Z., & Mojtaba, K. (2005). Characteristics of Fiber Bragg Grating Temperature Sensor at Elevated Temperatures. In *Proceedings of the 2005 International Conference on MEMS, NANO and Smart Systems (ICMENS'05).*

Chettouh, S., Akrmi, A. E., Triki, H., & Hamaizi, Y. (2017). Spectral properties of nonlinearly chirped fiber Bragg gratings for optical communications. *Optik, 147,* 163–169.

Daud, S., & Noorden, A. F. A. (2016). Fibre Bragg grating sensor system for temperature application. *Jurnal Teknologi, 78*(3), 39–42.

Daud, S., Aziz, M. S., Chaudhary, K. T., Bahadoran, M., & Ali, J. (2016). Sensitivity measurement of fibre Bragg grating sensor. *Jurnal Teknologi, 78*(3), 277–280.

Daud, S., Jalil, M. A., Najmee, M. S., Saktioto, S., Ali, J., & Yupapin, P. P. (2011). Development of FBG sensing system for outdoor temperature environment. *Procedia Engineering, 8,* 386–392.

Edmon, C., Chen, C. Y., Stephen, E. S., Stephen, J., & Ralph, P. T. (2004). Characterization of the response of fibre Bragg gratings fabricated in stress and geometrically induced high birefringence fibres to temperature and transverse load. *Institute of Physics Publishing, UK, 13,* 894–895.

Fasano, A., Woyessa, G., Janting, J., Rasmussen, H. K., & Bang, O. (2017). Solution-mediated annealing of polymer optical fiber Bragg gratings at room temperature. *IEEE Photonics Technology Letters, 29*(8), 687–690.

Fei, C. Y., Qun, H., & Liu, T. G. (2015). All-fiber optical modulator based on no-core fiber and magnetic fluid as cladding. *China Physics B, 24*(1), 014214.

Filho, E. S. D. L., Baiad, M. D., Gagné, M., & Kashyap, R. (2014). Fiber Bragg gratings for low-temperature measurement. *Optics Express, 22*(22)

Georges, T., Delevaque, E., Monerie, M., Lamouler, P., & Bayon, J. F. (1992). Pair induced quenching in erbium doped silicate fibers. *IEEE Optical Amplifiers and Their Applications, Technical Digest, 17,* 71.

Gouveia, C., Jorge, P. A. S., Baptista, J. M., & Frazoa, O. (2011). Temperature-independent curvature sensor using FBG classing modes based on a core misaligned splice. *IEEE Photonics Technology Letter, 23*(12), 804–806.

Grattan, K. T. V., & Meggit, B. T. (2000). *Optical fiber sensor: Advanced applications—Bragg gratings and distributed sensors.* Boston: Kluwer Academic Publishers.

Hill, K. O. (2002). *Fiber optics handbooks-fiber, deviced, and systems for optical communication.* McGraw-Hill.

© The Author(s) 2018
S. Daud and J. Ali, *Fibre Bragg Grating and No-Core Fibre Sensors,*
SpringerBriefs in Physics, https://doi.org/10.1007/978-3-319-90463-4

Hill, K. O., Fujii, Y., Johnson, D. C., & Kawasaki, B. S. (1978). Photosensitivity in optical fiber waveguides: Application to reflection filter fabrication. *Applied Physics Letter, 32*(10), 647–649.

Hirayama, N., & Sano, Y. (2000). Fiber Bragg grating temperature sensor for practical use. *ISA Transaction, 39,* 169–173.

Ho, S. P. (2008). *Properties of fabricated fiber Bragg grating for temperature sensing using phase mask technique.* Master Thesis, UTM.

Ho, S. P., Ali, J., Rahman, R. A., & Saktioto (2008). Growth dynamics and characteristics of fabricated fiber Bragg grating using phase mask method. *Workshop on Recent Advances of Low Dimensional Structures (WRA-LDSD).*

Huang, L. S., Lin, G. R., Fu, M. Y., Sheng, H. J., Sun, H. T., & Liu, W. F. (2000). A refractive-index fiber sensor by using no-core fibers (pp. 100–102). IEEE.

Jeff, H. (2002). *Understanding fiber optics (fourth edition)* (pp. 153–157). London: Prentice Hall.

Juergens, J. (2005). *Thermal evaluation of fiber Bragg gratings at extreme temperatures.* Canada Technology and Communication Research Center.

Jung, J. (1999). Fiber Bragg grating temperature sensor with controllable sensitivity. *Applied Optics, 38*(13), 2752–2754.

Kashyap, R. (1999). *Fiber Bragg gratings* (pp. 185–189). London: Academic Press Inc.

Kaur, G., Kaler, R. S., & Kwatra, N. (2017). Investigations on highly sensitive fiber Bragg gratings with different grating shapes for far field applications. *Optik, 131,* 483–489.

Kawasaki, B. S., Hill, K. O., Johnson, D. C., & Fuji, Y. (1978). Narrow-band Bragg reflectors in optical fibers. *Optics Letter, 3,* 66–68.

Krishna, V., Fan, C. H., & Longtin, J. P. (2000). Real-time precision concentration measurement for flowing liquid solutions. *Revision Science Instrument, 71,* 3864–3868.

Kumara, J., Prakash, O., Mahakuda, R., Agrawal, A. K., Dixit, S. K., Nakhe, S. V., et al. (2017). Wavelength independent chemical sensing using etched thermally regenerated FBG laser systems engineering. *Sensors and Actuators B, 244,* 54–60.

Lai, Y. C., Qiu, G. F., Zhang, W., Zhang, L., Bennion, I., & Granttan, K. T. V. (2002). Amplified spontaneous emission-based technique for simultaneous measurement of temperature and strain by combining active fiber with fiber gratings. *Review of Instrument, American Institute of Physics (AIP), 73*(9).

Lam, D. K. W., & Garside, B. K. (1981). Characterization of single-mode optical fiber filters. *Applied Optics, 20,* 440–445.

Li, C., Ning, T., Wen, X., Li, J., Zhang, C., & Zhang, C. B. (2015). Magnetic field and temperature sensor based on nano-core fiber combined with a fiber Bragg grating. *Optics & Laser Technology, 72,* 104–107.

Li, C., Ning, T., Zhang, C., Li, J., Wen, X., Pei, L., et al. (2016). Liquid level measurement based on a no-core fiber with temperature compensation using a fiber Bragg grating. *Sensors and Actuators A, 245,* 49–53.

Li, J., Zhang, W., Gao, S., Bai, Z., Wang, L., Liang, H., et al. (2014). Simultaneous force and temperature measurement using S fiber taper in fiber Bragg grating. *IEEE Photonics Technology Letters, 26*(3), 309.

Li, L., Xia, L., Wuang, Y., Ran, Y., Yang, C., & Liu, D. (2012). Novel NCF-FBG interferometer for simultaneous measurement of refractive index and temperature. *IEEE Photonics Technology Letters, 24*(24), 2268–2271.

Li, W., Cheng, H., Xia, M., & Yang, K. (2013). An experimental study of pH optical sensor using a section of no-core fiber. *Sensors and Actuators A, 199,* 260–264.

Lia, S., Li, X., Yang, J., Zhou, L., Che, X., & Binbin, L. (2016). Novel reflection-type optical fiber methane sensor based on a no-core fiber structure. *Materials Today: Proceedings, 3,* 439–442.

Lin, G. R., Fu, M. Y., Lee, C. L., & Liu, W. F. (2014). Dual-parameter sensor based on a no-core fiber and fiber Bragg grating. *OE Letters, 53*(5), 050502-1-3.

Liu, H. B., Hiu, H. Y., Peng, G. D., & Chu, P. L. (2013). Strain and temperature sensor using a combination of polymer and silica fibre Bragg gratings. *Optics Communications, 219,* 139–142.

Liu, Y., & Zhang, J. (2016). Model study of the influence of ambient temperature and installation types on surface temperature measurement by using a fiber Bragg grating sensor. *Sensors, 16,* 975.

Lopez, J. E. A., Mondragon, J. J. S., LiKamWa, P., & Arrioja, D. A. M. (2004). Fiber-optic sensor for liquid level measurement. *Optic Letter, 36*(17), 3425–3427.

Malo, B. (1995). Apodised in fiber Bragg grating reflectors photoimprinted using a phase mask. *Electronic Letters, 31,* 223–225.

Meltz, G., Morey, W. W., & Glen, W. H. (1989). Formation of Bragg gratings in optical fibre by transverse holographic method. *Optics Letters, 14*(15), 823–825.

Miao, Y. P., Ma, X., He, Y., Zhang, H., Zhang, H., Song, B., et al. (2016). Low-temperature-sensitive relative humidity sensor based on tapered square no-core fiber coated with SiO2 nanoparticles. *Optical Fiber Technology, 29,* 59–64.

Michael, K., Simon, K., Johanne, H., Christian, B., Axel, S. H., Frank, V., et al. (2014). Fabrication and characterization of Bragg gratings in a graded-index perfluorinated polymer optical fiber. *Procedia Technology, 15,* 138–146.

Morey, W. M., Ball, G. A., & Meltz, G. (1994). Photoinduced Bragg gratings in optical fibers. *Optics and Photonics News,* 8–14.

Muhs, J. D. (2007). Design and analysis of hybrid solar lighting and full spectrum solar energy system. *Oak Ridge National Laboratory.*

Mulle, M., Wafai, H., Yudhanto, A., Lubineau, G., Yaldiz, R., Schijve, W., et al. (2016). Process monitoring of glass reinforced polypropylene laminates using fiber Bragg gratings. *Composites Science and Technology, 123,* 143–150.

Neil, J. G. (1999). Development of temperature compensated fiber optic strain sensors based on fiber Bragg gratings. University of Toronto.

Othonos, A. (1997). Fiber Bragg gratings. *Review of Scientific Instruments, 68,* 4309–4341.

Othonos, A., & Kalli, K. (1999). *Fiber Bragg gratings: Fundamentals and applications in telecommunication and sensing.* Boston (USA): Artech House Inc.

Park, J., Kwon, Y. S., Ko, M. O., & Jeon, M. Y. (2017). Dynamic Fiber bragg grating strain sensor interrogation with real-time measurement. *Optical Fiber Technology, 38,* 147–153.

Pedersen, J. K. M., Woyessaa, G., Nielsena, K., & Bang, O. (2017). Effects of pre-strain on the intrinsic pressure sensitivity of polymer optical fiber Bragg gratings. In *Proceedings of SPIE.*

Rahman, R. A., Ikhsan, S., & Supian, H. M. (2000). Fibre optic Bragg grating sensors: A new technology for smart structure monitoring in Malaysia. IEEE. 0-7803-6355-8.

Ramesh, S. K., Kuo, C. W., Righini, G. C., & Najafi, S. I. (1999). Design and fabrication of a Fiber Bragg Grating Temperature Sensor. *International Society for Optical Engineering Proceedings Series.*

Ran, Y., Xia, L., Han, Y., Li, W., Rohollahnejad, J., Wen, Y., et al. (2015). Vibration fiber sensors based on SM-NC-SM fiber structure. *IEEE Photonics Journal, 7*(2), 6800607.

Rezayat, A., De Pauw, B., Lamberti, A., El-Kafafy, M., Nassiri, V., Ertveldt, J., et al. (2016). Reconstruction of impacts on a composite plate using Fiber Bragg Gratings (FBG) and inverse methods. *Composite Structures, 149,* 1–10.

Ronald, S. (2003). Seeing the light. Trade & Industry.

Shilton, A. N., Powell, N., Mara, D. D., & Craggs, R. (2008). Solar-powered aeration and disinfection, anaerobic co-digestion, biological CO(2) scrubbing and biofuel production: The energy and carbon management opportunities of waste stabilisation ponds. *Water Science Technology, 58*(1), 253–258.

Simpson, A. G. (2005). Optical fibre sensor and their application. Ph.D. Thesis, Aston University.

Strasser, T. A., Pedrazzai, J. R., & Andrejio, M. J. (1997). Reflective mode conversion with UV-induced phase grating in two-mode fiber. In *Conference on Optical Fiber Communication (OFC) 1997* (pp. 16–21). Dallas, TX: FRS.

Snyder, A. W., & Love, J. D. (1983). *Optical waveguide theory*. London: Chapman & Hall.

Tanaka, T. (2002). 100 GHz Spacing 8-channel light source integrated with gratings and LDs on PLC platform. Technical Report of IEICE.

Tahir, B. A., Ali, J., & Rahman, R. A. (2005). Strain measurements using fibre Bragg grating sensor. *American Journal of Applied Science (Special Issue)*, 40–48.

Song, Z. J., Zhong, H. Q., Wei, L. X., Wu, Z. M., & Xia, G. Q. (2015). Experimental investigations on nonlinear dynamics of a semiconductor laser subject to optical injection and fiber Bragg grating feedback. *Optics Communications, 354*, 213–217.

Su, G., Shi, J., Xu, D., Zhang, H., Xu, W., Wang, Y., Feng, J., & Yao, J. (2016). Simultaneous magnetic field and temperature measurement based on no-core fiber coated with magnetic fluid. *IEEE Sensors*, 1–5.

Udd, E., Nelson, D., & Lawrence, C. (1997). Multiple axis strain sensing using fibre gratings written onto birefrigent single mode optical fibre. *Optical Fibre Sensors*, 354–357.

Wang, X., Dong, X., Zhou, Y., Ni, K., Cheng, J., & Chen, Z. (2013). Hot-wire anemometer based on silver-coated fiber Bragg grating assisted by no-core fiber. *IEEE Photonics Technology Letters, 25*(24), 2458–2461.

Wang, Y., & Wen, Y. (2016). Temperature and strain sensing properties of the zinc coated. *Optik, 127*, 6463–6469.

Woyessa, G., Pedersen, J. K. M., Fasano, A., Nielsen, K., Markos, C., Rasmussen, H. K., & Bang, O. (2017). Simultaneous measurement of temperature and humidity with microstructured polymer optical fiber Bragg gratings. SPIE. 10323. 103234T-1.

Xia, L., Li, L., Li, W., Kou, T., & Liu, D. (2013). Novel optical fiber humidity sensor based on a no-core Fiber structure. *Sensors and Actuators A, 190*, 1–5.

Xiaopei, C., Yan, Z., Gang, P., & Jiju, A. (2004). Experimental design in fiber optic development. *International Journal of Productivity and Performance, 53*, 8.

Xie, J., Li, H., Gao, L., & Liu, M. (2017). Laboratory investigation of rutting performance for multilayer pavement with fiber Bragg gratings. *Construction and Building Materials, 154*, 331–339.

Yang, H. Z., Ali, M. M., Islam, M. R., Lim, K. S., Gunawardena, D. S., & Ahmad, H. (2015). Cladless few mode fiber grating sensor for simultaneous refractive index and temperature measurement. *Sensors and Actuators A, 228*, 62–68.

Yao, Q., Meng, H., Wang, W., Xue, H., Xiong, R., Huang, B., et al. (2014). simultaneous measurement of refractive index and temperature based on a core-offset Mach-Zehnder interferometer combined with a fiber Bragg grating. *Sensors and Actuators A, 209*, 73–77.

Yonghang, S., Jinglei, H., Weizhong, Z., Tong, S., Kenneth, T. V. G., & William, D. N. P. (2004). Fiber-optic sensor system for heat-flux measurement. *Review of Scientific Instruments*.

Yua, H. C., Fan, Z. Y., Xia, Z. M., Gordone, L. L. M., & Qiang, L. L. (2016). Application of FBG sensors for geotechnical health monitoring, a review of sensor design, implementation methods and packaging techniques. *Sensors and Actuators A, 244*, 184–197.

Zhang, C., Xu, S., Zhao, J., Li, H., Bai, H., & Miao, C. (2017). Intensity-modulated refractive index sensor with anti-light source fluctuation based on no-core fiber filter. *Optics and Laser Technology, 97*, 358–363.

Zhang, R., Man, J., Liu, W., Zhou, W., & Yang, Y. (2015). A multi-band access radio-over-fiber link with SSB optical millimeter-wave signals based on optical carrier suppression modulation. *Optical Switching and Networking, 18*, 235–241.

Zhou, G., Wu, Q., Kumar, R., Ng, W. P., Liu, H., Niu, L. F., et al. (2017). High sensitivity refractometer based on reflective SMF-small diameter no core fiber structure. *Sensors, 17* (1415), 1–9.

Zhang, C., Xu, C., Zhao, J., Li, H., Bai, H., & Miao, C. (2017). Intensity-modulated refractive index sensor with anti-light source fluctuation based on no-core fiber filter. *Optics and Laser Technology, 97*, 358–363.

Zhang, W., Ying, Z., Yuan, S., & Tong, Z. (2015). A fiber laser sensor for liquid level and temperature based on two taper structures and fiber Bragg grating. *Optics Communications, 342,* 243–246.

Zhao, J., Wang, J., Zhang, C., Guo, C., Bai, H., Xu, W., et al. (2016). Refractive index fiber laser sensor by using tunable filter based on no-core fiber. *IEEE Photonics Journal, 8*(5), 6805008.

Zhou, G., Wu, Q., Kumar, R., Ng, W. P., Liu, H., Niu, L., et al. (2017). High sensitivity refractometer based on reflective SMF-small diameter no core fiber structure. *Sensors, 17* (1415), 1–9.

Zibaii, M. I., Latifi, H., Karamia, F., Ronaghi, A., Nejad, S. C., & Dargahi, L. (2005). In Vivo brain temperature measurements based on fiber optic Bragg grating. In *Proceedings of SPIE.* 10323. 103234G-1.

Index

© The Author(s) 2018
S. Daud and J. Ali, *Fibre Bragg Grating and No-Core Fibre Sensors*, SpringerBriefs in Physics, https://doi.org/10.1007/978-3-319-90463-4